I0484210

SILVIBA

A forester's log

Tony Richmond

Copyright © 2015. All rights reserved.

This book is not-for-profit.

Net proceeds are dedicated to the Prince Albert Women's shelter.

C/O Carol Soles, 306-764-7233

Most names in this narration are factual. A few are changed to deflect from sensitive issues dealt with from time to time.

Ω

The photo on the front cover: Sandy Ross on his trap line at Two Forks, Saskatchewan. Note his 'value-added' load of birch bark. He was a treaty Indian and a member of Montreal Lake Cree Nation in Saskatchewan. He is buried at Molonosa, the site of the signing of the adhesion to Treaty 6 in 1889. Used with permission.

Ω

ISBN: 978-1508989189

SILVIBA

A forester's log

Ω

(A narrative nonfiction account of a forester's career in Western Canada)

Anthony Richmond was born under the walls of Dartmoor Prison in Devonshire, UK. He immigrated to Canada in 1949, and received his Canadian naturalisation at New Westminster in 1957. He graduated from UBC in 1958, received his professional forester's registration (RPF no. 403) in 1963 and thrived in the logging camps of the Nimpkish Valley on Northern Vancouver Island. He married Marilyn Elizabeth Livingston, a teacher at Nimpkish camp. His career took him from industrial forestry to university lecturer, to an instructor at BCIT after achieving a Master's degree in Forest Management. In the process, he participated in the formation of International Forest Fire Systems Inc. After a stint in Prince George for Industrial Forestry Services Ltd he ended in Saskatchewan as the government's chief forester. He developed new forest policy as part of his terms of reference. This got him fired. He started up Silviba Services Ltd at Regina in 1984, taught at Lakehead University, and then opened a forestry services office in Prince Albert, SK.

He spent the next thirty years working with small sawmill operators, Indian bands with forestry aspirations, a campaign to rid Saskatchewan of the Forestry Crown corporations, timber salvage from forest fires, and research into the early forest history of the Prince Albert district, NWT. His most absorbing task was the reconstruction of forests on certain Indian reserves to obtain estimates of the volume of stolen timber as part of Indian Special Claims timber claims – Forensic Forestry.

Ω

Visit the Silviba website at www.silviba.com

Dedication

Marilyn E. Richmond, who provides a rock-solid platform for me at all times. My heartmate. We raised a resilient family by necessity. Without my wife, Marilyn, I could never have been the forester I was supposed to be. Without my children and grandchildren, I would not be the whole person I wish to be. Marilyn is within me in spirit on all my journeys.

Wendy (Wiebechick) Wiebe, Silviba's essential service and soulmate.

To all the forestry spouses who followed their partners into the more remote reaches of the Canadian forest landscape to make a home.

Ω

SILVIBA

This is pronounced 'Sil-V- 'eye' -Ba'. Meaning 'silviculture' (growing trees) and 'fibre' mowing them down – logging.). Sustainable forestry.

Thank you

Thank you to Saskatchewan Archives Board and the Library, University of Saskatchewan, Prince Albert Museum.

Ω

Thank you to my book-ends: Elspeth Richmond, Kathy Richmond, Melissa Richmond, my agents – Gillian Kydd and Nancy Gardner. To Bob Romanchuk, and my shepherdess – Suzanne Doyle-Ingram (www.suzannedoyleingram.com); also to my readers – Kim Fenton and Annie Grant. To my reviewer, Cathie Roy. Lastly, to Nancy Gardner, who was as close to an editor as you can get.

Ω

Thank you to my Canadian Indian connections at Montreal Lake, Pelican Narrows, Southend, Sturgeon Landing, Sandy Bay, Deschambault, La Ronge, Weyakwin, Two Forks, Pinehouse, Canoe Lake, Prince Albert, Meadow Lake, Buffalo Narrows, Joseph Bighead, Dakota Whitecap First Nation, Mistawasis, Big River, Muskoday, Sandy Lake, Ontario, Moose Lake, Manitoba – Chiefs, councilors and band members. And yes, to Mary Ballantyne who ran the Deschambault forestry office.

Ω

Gratitude

Roy Jewesson, my first mentor: he allowed me to do my job with a minimum of intrusion. I excelled in this atmosphere.

Patrick Conway, who taught me to log highlead: The most underrated performer on the Englewood payroll.

Owen Hennigar, the divisional manager, Englewood logging division that produced four million plus cubic metres of logs in a year.

Harry Gairns, Prime Mover at Industrial Forestry Services Ltd, who allowed me to develop my forestry technical skills, participate in the development of a strategy to allow students in forest technology to proceed into university, and then end up with a professional forester designation.

Bob Romanchuk and Uncle Harry, together they kept my ship afloat after I was thrown overboard by the government of Saskatchewan.

Dave Lindenas, the catalyst for Silviba's forest inventory data acquisition, from both his government and Weyerhaeuser forest inventory positions.

Gene Kimbley, who enabled me to retool our GIS apparatus with money from the Aboriginal Business Development Program. And, who brought me back to Jesus.

Gertie Budd, councilor, Peter Ballantyne Cree Nation, Sturgeon Landing, a pleasure to work for.

Jackie Ross, First Nations, trapper, logger, and Hummer advocate who provided all the cost data used to construct logging proformas within many First Nations forestry enterprises business plans. Jackie is a true friend, colleague, fellow conspirator and collaborator. He lives on his trapline at Two Forks, Saskatchewan (SK).

British Columbia Institute of Technology (BCIT), the institute where I was encouraged to develop a forest fire simulator training program for use with students, industry, universities and governments, as well as allow me to help develop International Forest Fire Systems Inc. (IFFS), and who allowed me to test innovative teaching protocols.

The Faculty of Commerce Graduate Studies in 1969 at the University of British Columbia (UBC), where I studied 'Organisational Behaviour and Administration'. The understanding and education I gained there changed my life.

Alphonse Bird: First Nations, councilor, chief and Grand Chief, from his homeland community of Montreal Lake, SK.

Jack Custer, First Nations councilor, a good friend, from his homeland community of Deschambault Lake, SK. I miss you, Jack.

Henry Morin, First Nations, 'Mr. per diem', councilor, wannabe chief, perennial champion of economic development project attempts, from his homeland community of Pelican Narrows, SK.

Jason Nelson, who taught me the role of soils in association with forest ecosystems.

Carrier Lumber, Kordybans *et a*l, a treasured connection.

Dr. Thomas Bouman, Prince Albert Model Forest's first director. And so many more.

Ω

Table of contents

Table of figures

PREFACE

The frame for this journey encompasses my experience as a professional forester within various settings that span some fifty-five years in length. Why set down an account of this journey? First, it is difficult for me to define concisely the job description of a professional forester without exploring my experiences. Many times in my career, folk have asked me this question: what do you do? I felt inadequate in my responses. You will need to complete the journey to earn a more focused understanding. Part Two includes an account of the Prince Albert Model Forest, one of ten set up across Canada in 1992. The federal government set up the program in association with each participating province to offset international criticism for the absence of environmental stability within Canada's forests.

In Part Three, I relate the research and results of large-scale theft of timber from Indian reserves immediately north of Prince Albert. In Part Four I make a case for a new ethnic label: *'Candians'*, with a subset, *First Candians*, rather than just 'Indians'.

Second, I remain perplexed over the absence of progress in the association of Indian and non-Indian agendas after thirty years involved with Saskatchewan forestry. We bring in Chinese to operate a mine, and many more through our immigration program to seed additional labour supply. However, we abandon Canadian Indians to a sinkhole of poverty, and often despair by pirating their landbase. So, my eventual mission in Saskatchewan was to apply my skills and experience to foster socio-economic gain within the Indian world using the forest available. I have failed miserably. However, after I present some case histories recording my Indian involvement perhaps you will identify ingredients leading to a more osmotic approach to the legitimate connection for Indians within our larger society. Although, it does not help when Indians, mainly on reserve, avoid association, and the rest of us are sublimely indifferent. The cultural wreckage caused by the Indian residential school era will be with us for at least three more generations. Neither does it help to watch the white man creating an economy built on Indian lands and resources, but expects the Indian world to follow along. Those Indian treaties signed so long ago did nothing but create dependences that have

xix

grown exponentially over time. Part Three paints a picture of the early forest history in the Saskatchewan district, NWT. The Indians got screwed. In Part Four, I offer conclusions and recommendations with respect to forest administration, *Candians* and the creation of sub-treaties, and the impact of the banking industry on Silviba. Carbon capture assessments are needed as well. This becomes the endgame for the journey.

Lastly, the reader will need access to Google and Google Earth. Foresters live and die for their maps. To follow the thread of the narrative to come, you may need to follow up the website links.

Despair not, the Angel declared, for joy is always within. When this meets up with joy without, your blessings sing out.

Anon.

Ω

PART ONE
British Columbia (BC)

Chapter 1:

Beginnings

The grey leaden sea erupted high above the rear deck, thrusting up the ship's stern, before it slipped down into the next trough. My job was to exercise Judy, the family black Standard poodle, on that heaving bottom deck. I do not remember entering the ship's restaurant once on that trip to Canada from Liverpool. I survived on what little Canada Dry I could keep down. Here was I, a pimply sixteen year-old with my own British passport. The family sailed in the Empress of France across the Atlantic while plowing into the teeth of Atlantic Hurricane 13, causing a two-day delay. This was the last year in which the authorities numbered hurricanes. From then on, they had names. We sailed up the St Lawrence River past Quebec City and finally docked in Montreal on October 25, 1949. By midnight, the family was tucked in on the sleeping car of a Canadian Pacific Railway immigrant special parked at Windsor station across from the docks. After nine days of vomiting I did not notice the absence of a restaurant car. Poor Judy found herself in with the baggage. After stops for snacks at White River in northern Ontario, a stop for snacks in Winnipeg, then a long haul across the prairies, I was mesmerised by the scenery between Winnipeg and Calgary. It was almost November – no snow yet – but after a full day with my forehead glued to the window I could absorb no more. Anyway, it was getting dark. And, I finished up with a roaring headache.

The train arrived in Calgary just after a heating breakdown – the temperature shot up and gave an American gentleman a heart attack. The family discovered tea-bag tea at that time as we sat and felt the heat rise in the little section reserved for afternoon tea. It was getting dark as the train left Calgary. I woke up at the top end of the Fraser Canyon at Lytton on the fourth morning.

Dad had a job to go to in Vancouver. Everyone camped in an old house at Third street and Third Avenue in New Westminster. Canada welcomed me in a manner that was pure Canuck – the train whistle blowing sometime in the night as it crossed the Pattullo Bridge. The sound bounced off structure after structure on my side of the Fraser River. I enrolled in grade 12 classes at Duke of Connaught high school. In January, the class was to research possible career paths. I chose 'forestry'. Don't ask me why. On the shelf, I found one pamphlet on a career as a British Columbia log scaler put out by the BC Forest Service, nothing more in the forestry line. The rest of the forestry shelf was empty. This, in British Columbia in which the major industry was logging, sawmills and pulpmills? Since I still dressed 'English', I spent the whole time at school defending my tie from the world around me until a false graduation next June. My diploma was blank.

I then spent the next two years going to night school and completing grade 12 courses while working at Stubby soft drinks in New West. I went on from there to become a clerk in the BC Forest Service log scaling office in the basement of the Marine building in downtown Vancouver. There were six women and me. They cranked out invoices for Crown dues originating from scale sheets provided by ministry log scalers from every log boom tied up in every bay and backwater on the British Columbia coast. My job was to check that each invoice produced matched the scale sheet summaries and the Crown dues billed. Then, before each invoice went out, I had to stamp the pink copy and initial it as correct. On one occasion, I let an invoice to Alaska Pine and Cellulose for $4 million-odd get out to the mail. This company had a heart attack on receiving the bill. The corrected invoice was for $40,000 or so. I received an Imperial invitation to visit the Chief Clerk, Eric Fox, twelve floors up. It took a few weeks to live that down. On another day, I received a phone call from Eric Fox on another matter. Unfortunately, before I realized who it was, I answered the call by playing a tune from a musical instrument. I had rigged up elastic bands stretched at various tensions across the pigeonholes at my desk. I made some stupid explanation the chief clerk received gracefully.

I must mention the six women that produced the invoices. There had been no girls in my school life – just matron and Miss King, who taught us English, so, having grown up in a world embedded with boys, I coped as best as I could. Jen Stoddart, our supervisor, operated from a wheelchair. She could have been my mother. Katherine Terard, second in command and a confirmed spinster, acted as an aunt. She served as a surrogate for my own Aunt Doris, who had brought us up in the War. Ruth Donnelly, who was about to get married, had exquisite breasts. I met them at one picnic on Wreck Beach when I stole a look down her blouse. The last two women were Maureen and Rachel Pyke. The former was my age, and had frizzy, long hair. You know that period in a girl's life just before she discovers haircuts and transforms herself into a woman? One evening she invited me for supper at her folk's house on a float in False Creek – before the haircut. The latter was a young wife of a navy doctor posted overseas. She was blond and well turned out. She had on a few occasions attempted to separate me from the pack. This time, she generated the idea of an after-work horse ride in Stanley Park, just her and me. Well, one day we set out after work to do just this. I rode my horse right around Stanley Park – just me. Her horse had refused to leave the stable.

Shortly after that, I found myself enrolled at the University of British Columbia in first year Arts and Science (September 1952). I played freshmen rugby that winter. On one wet November day, we played a game against the Meralomas on the Kitsilano playing field. That team was huge. I played left wing. My opposite number picked me up by the seat of my pants and threw me into the stands after one tackle. Then it was my turn. As this lumbering elephant trundled towards me carrying the ball, I went into a crouch. As he jumped over me, I raised my hand and latched on to his ankle. He came down like a sack of spuds. Unfortunately, I had to work weekends after first year. No more rugby. The following spring, after the end of my first year, I turned up at the Green Timbers Forest Nursery in Surrey, together with the rest of the Forest Service summer student contingent.

Ω

Chapter 2:

Summer 1953

That spring, I joined Mickey Pogue's army and became part of the Kaslo contingent housed in the garage at the BC Forest Service depot. Mickey was the forester in charge of forest inventory for the BC Forest Service. The federal government had combined with each forestry province to undertake an up-to-date forest resource inventory after the end of World War Two. He designed a lightweight sleeping bag that came up to one's nipples. On the way up we stopped for a break in Penticton. We were to meet back at a certain pub before continuing the trip to Kaslo. The bartender kicked me out before I had a chance to rejoin them so I squatted on the outside steps to wait. We travelled in an Austin 7 station wagon. A governor on the engine set the throttle limit at forty mph. We burned out that governor halfway to Nelson. En route, we passed the government minister responsible for the BC Forest Service. He reported us to our boss in Victoria and turned us in for speeding – the Forest Service decals gave us away. This happened on the U.S side of the border. The boss restricted us from driving through the Canadian Monashee Mountains – too many drop-offs. Extra bodies joined the project in Nelson, plus a riverboat. I rode in the back of the riverboat on a trailer from Nelson to Kaslo. We crossed the Kootenay River on the ferry to get to the north shore. Each time I looked over the edge, on our way to Kaslo, I was staring down into the top of a vertical drop ending in Kootenay Lake. The job entailed the sampling of forest populations in the north part of Kootenay Lake, using Kaslo as a base. The crew had riverboats for transport and dried rations for a week. We were never out for less than ten days. Backpacks weighed ninety pounds when you added in all the survey equipment used to collect sample ages, heights and diameters for each tree species present. The pack also contained a week's supply of freeze-dried food. We added in a bar of dark Baker's chocolate. No helicopters then. In crews of two, we sampled trees from the bottom of each valley up to the treeline in those West Kootenay creeks located in the Selkirk Mountains.

Forest modeling as a forest management tool

Our job was to supply samples of the forest inventory data collected from labelled forest types. The purpose was to build a forest database and create forest inventory maps back at Forest Service headquarters in Victoria. The air photos arrived in Kaslo, pretyped[1]. Mickey Pogue's forest interpreters in Victoria had drawn boundaries around each type (polygon[2]) in each landscape parcel from the Lake edge, all the way up to the alpine. They grouped stands of timber with similar characteristics – timber species, crown density, height class, aspect, elevation and other identifiers. They did this by viewing each aerial photo along selected flight lines using stereoscopes. The images stood up so an interpreter could label the forest cover within each polygon according to the tree species identified. Next, the interpreters subdivided these polygons by tree height and terrain elevation. This is an essential ingredient in forest modeling. The one tree feature impossible to obtain from a stereo photo image is the age of a tree. However, a graph, using our plot data, provided the relationship between tree height and its age. This produced a value for total age. Those air photos were in much disrepair when returned to Victoria. Pinpricks through each photo recorded the location and timber type for each sample plot. Squashed mosquitos found their way to Victoria embedded on the backs of these badly mauled air photos. Often, we disagreed with the first pretyping label. After all, we viewed the forest firsthand while the interpreter only had an aerial photo at a scale of 1 inch on the photo that equaled 31,680 inches on the ground, or 1 cm on the photo equaled 805 metres on the ground. A significant handicap in photo interpretation.

Total age of each tree measured in each sample plot then links to its diameter-at-breast-height (dbh). This, in turn, provides the growth rate for that polygon. A second connection links age and diameter – all from the same plot data. Lastly, a relationship develops between timber volume by tree or by acre (hectare) and the age of that polygon. This produces a sigmoidal curve (an S curve) defining volume over age for each tree species sampled. The curve rises from zero, peaks at a biological maximum (rotation age) before retreating slowly down to old age.

Hanzlik[3] incorporates the notion that we only harvest timber at the biological rotation age. The timber volume removed in any year must not exceed the annual volume (biomass) replaced by growth. In simple terms, this is the Annual Allowable Cut, or sustained yield. The Saskatchewan equivalent was the Harvest Volume Schedule. One, the allowable cut should preserve the proportion of tree species within the forest planning unit. Two, exceptions develop from the occurrence of forest fires, insect attacks, blowdown or flood. The forest authorities will issue an instruction to salvage this timber before switching back to green timber.

In practice, there are a number of rotation ages in play. One is the economic rotation. Trees younger than biological rotation age are harvested because market values have peaked, or, the sawmill requires a preferred log size. The demand for poles and piling come to mind. A stand of trees just up from where I now live was recently logged for Douglas fir poles. Ring counts on stumps ranged from 98 to 120 years. The stumps reached three feet in diameter. Tree heights were 40 metres or so. Next to this logged area is the remnant of a pure Douglas fir ecotype on the Sunshine Coast. BC Timber Sales[4] has marked it for logging. Local citizens are fighting for its survival. That stand of timber is too young to fell and represents the last remnant of a pure Douglas fir forest ecotype on the entire Sunshine Coast.

Another rotation age is one of expediency. A stand of trees perhaps is dying from a contagious root disease. Or, perhaps insects are killing a stand of younger trees, so salvage becomes a priority.

Gifford Pinchot became The U.S Forest Service's first chief forester in 1905. He had just finished a year of training in France. He brought back the Hanzlik formula for calculating allowable cuts. This originated in Europe in the 1800s. It incorporated the notion that the biomass logged each year from a forest district should not exceed its sustainable capacity. In British Columbia, legislation gave birth to the Forest Service in 1912, under the chief forester, M.A.Grainger.

(BC Forest Service history.
http://members.shaw.ca/wolfpatch/Forestry/History/).

Ω

Before the first survey trip we all gathered down at the mouth of the Kaslo River. There was a large gravel bar supporting a clump of cottonwood trees that had survived flood, tempest and Lord knows what else. We took the measurements of tree diameters at breast height over bark, and then total height using an Abney, also total age – oh yes! Those cottonwoods were spongy in the middle. The increment borer, once embedded the necessary distance, got stuck, and would not spiral its way back out. This was a forestry assistant's nightmare. How to extract the inky when it would not twist its way back out? Or, when the core extractor got stuck down the barrel of the inky shaft? You counted the tree rings on the core extracted from the tree. This provides the age at breast height. Note that each annual ring is a sandwich of light spring wood between bands of dark summer wood – one dark band per year. You use a table of height-over-age values to get the years to grow from ground level to the height at dbh).[5] Add the two numbers, and you have total age. Those cottonwoods also had a further surprise. As I pulled the extractor out to count the annual rings, a stream of the most evil smelling liquid shot out and covered my front. My mates told me that I smelled like a moose in the rut. This whole aging thing was a constant struggle. Sometimes it took two of us to reef on the inky to twist it out of a tree. It was Swedish-made of high quality steel. The auger end had an external cutting edge around the hollow barrel. The metal was sharp but tapered as it came to a point at the extreme tip. Many an auger was chipped at the end. The last hazard was a bent extractor. Taking the age of a standing tree is problematic at times. Our sampling locations were many, many miles from home, and up some steep mountainsides. During my career, I had to chop the inky out of the tree a few times. By the way, have you ever sat down to count the growth rings extracted from a sample tree? Add in hot humid summer weather, mosquitoes, mosquito repellant mixing with sweat to irritate your eyes, and dive-bombing deer flies. Purgatory.

That first tramp into the forest

That first tramp into the forest occurred along a creek above Ainsworth, close to our base and accessed by truck. My partner, Dick Greyson, was on his second year in Mickey Pogue's army. The previous summer he had helped to extract a forest lookout man, who was badly torn up by a grizzly bear, from the fire lookout at Beaton. This was just northwest of the top of Kootenay Lake. Dick's dad ran a pharmacy in Nelson. Dick was paranoid over the chance of meeting a big grizzly when he and I were partners. We tied metal cups to our Trapper Nelson backpacks, yelled out regularly and were careful where we stashed food overnight.

Early May, in four feet of wet snow under a canopy of veteran hemlock-cedar, we followed Ainsworth Creek upwards all day. No snowshoes, snow up to the crotch, 90-pound backpacks. Up and up through soaking wet hemlock and cedar forests. Somewhere under that snowpack was a roaring creek full of water and nests of hidden windfalls. Occasionally, one foot would slip into a hole. There I would be – my pack pressing down while I attempted to free my foot and get back up on to something stable while preserving the family jewels.

The destination was a patch of spruce-balsam below the Kokanee glacier. Above us, the bowl below the glacier, an open and barren snowfield, served as a back door. That evening the temperature fell to below zero. We set up camp by raising the fly sheet over a low spot in the timber, cut spruce boughs to serve as a bed, cooked the dried rations after melting snow, and slept. Dick taught me to be comfortable. That was how I spent my first night ever in the Canadian wilderness.

The next morning we set up a sample position using the air photos, and took tree diameters, total age and sample tree heights for the dominant trees in the canopy. I had a creepy experience while preparing to take a tree height. You see, to take a height, you aimed the Abney at the bottom, and then you walked out two chains[6] from the base of the sample tree then obtained a top reading. In this case, I backed up until the chain said "2", and then looked back. I was slap in front of a small cave where the occupant had been busy. Muddy tracks in the snow lead into and away from the entrance. With my rear to the lair and the hair on my back standing straight up I took the Abney shots, got the tree height measurements and scuttled away. Wolverines, when cornered, are vicious. After some further education, I learned to recognise their tracks and give them a wide berth. Those were wolverine tracks at that small cave. I hate taking tree heights. The line of sight between me and the tree is always blocked by intruding vegetation and tree trunks. The only remedy is to have someone hold a flashlight at breast height on the target tree. This reduces the guessing game immensely.

We then spent the rest of the day wading home in the snow until we reached the road. At least it was all downhill. I realize now that this was a picnic compared with the trips to follow.

The next occasion plus
We arrived at the mouth of Hamill Creek using the riverboat, just across Kootenay Lake, and opposite Lardeau. The trail ran close to the creek and was an old raw-hiders' route. In fact, it was one of the proposed crossing points for the Trans-Canada Highway, linking the West Kootenay with the East Kootenay. Eventually, of course, the Rogers Pass became the preferred alignment for this highway. We plodded up this old overgrown mining trail littered with windfalls. I thought of those raw-hiders with their mules as I struggled across each windfall, my pack pulling me back. On arrival at the base of a big scree slope rising a thousand feet at least, bears from the upper slopes moved down to get our scent since they do not see too well over large distances. There were black bears, brown bears and grizzlies coming down that slope. Given the season was still early, the skunk cabbage and fescues were just beginning to appear. The bears, just recently emerging from hibernation, were hungry. However, as they caught our scent, they wandered back uphill. Given my innocent attitude to all the animals wandering around, I missed some of the excitement – except for one nighttime. Dick and I camped in an old wickyup. We were way up Hamill Creek in an old-growth stand of cedar-hemlock. I slept beside a big windfall and left my Heads Light Cruisers (leather caulk boots, brand-new) on the tree to dry out above the foot of my sleeping bag. I woke to the rustling sound of something moving around at my feet. I was scared stiff. To make matters worse, I felt a sudden weight fall on my feet. So I froze in terror. It stayed on

my legs until morning since I dared not move. Of course, it was one of my boots that had fallen – or was pushed – off the log. We had heard stories of crews that had their footwear eaten by porcupines who went after the salt in the leather, and so we were especially careful to guard against this happening. The highlight of that particular location was that Dick found a large old-growth cedar with some bark removed with an axe. On the bare wood someone had written in pencil his name and the date – 1916. Dick recognised the name as belonging to a man in Nelson. As a sidebar to this bear stuff, I developed a theory early on that bears rush out to search for ants following hibernation. The formic acid from ants gets the bears' digestion started back up. I often see signs of disturbed forest debris along forest trails.

One time a grizzly, whose tracks in the snow were as large as your largest dinner plate, followed us. We stepped in its tracks on the way back. On my trek back to the boat after ten days up Hamill Creek, I was walking along this windfall hemlock minding my own business. As I jumped down off the end, I caught sight of a grizzly sleeping in a spot of sunshine close to where I was to land. Without pausing, I hurtled myself over it and raced a mile to the creek. Melting snow from up on high had converted the creek into a rushing torrent – a freshet. Dick and I knew that we had to cross to put distance between the grizzly and us. Once there, Dick sized up the current and developed a plan to get us across alive. He was to pick up a long pole, walk along a thrumming tree sticking into the stream, then jump and land on the other side. This he did, except that he was swept downstream immediately and around a bend. He disappeared for ten minutes. He showed up on the creek bank across from me and indicated that I should grab on to another long pole he was holding firmly out over the creek. So I crept along the thrumming tree as the water washed over my feet in waves, grabbed the pole and jumped while wearing my 90-pound backpack. Yes, I also was swept downstream around the bend and came to the other shore eventually. I crawled out of the water and rejoined Dick. No more preoccupations with grizzlies on that occasion. We were probably the first and last folk to "swim" across Hamill Creek during a May freshet.

It was a tough trip – the constant rubbing of Dick's sock against his ankle had created a deep infection. He became really worried that blood poisoning would settle in. This impacted on me since I would have to pack him out. So, I started to worry as well. And we had run out of food. When we arrived down at Argenta, a Quaker community newly moved in from California, someone pulled milk up from the well and fed us real food for the first time in ten days.

Ω

On June 2, 1953, Dick and I sat sunbathing on a beautiful beach up from Johnson's Landing as we waited for a boat to take us back to base camp in Kaslo. It was Coronation Day for Her Majesty Queen Elizabeth II of Great Britain. The only reason I knew this was that I had a tin of sweets with her picture on it, and her coronation date.

At some stage, back at base, everyone decided it would be a good idea to take the doorknob off Bob's door while he was still inside the room. Bob was the project leader and a forester from South Carolina. He used the room as his office and sleeping quarters. Anyway, when he found he could not open his door he let forth with orders, exhortations and demands. He tried to get out of his window, but it was a high, narrow one that opened from the top and swung out. The instigators replied: 'We don't have your little old dooooorknob, Bob', then everyone left for the pub except me. I reinstalled Bob's doorknob and listened to the complaints. Ted Lavis replaced him soon after.

Later that summer up the Duncan River I sliced through my beautiful Heads light cruisers and cut off a toe – almost. Seemingly, my renown for the sharpness of my axe caught up with me. I had almost severed the toe, but the doctor, newly immigrated from Spain, sewed it back on using one last flap of skin (without anesthetic and with lots of brandy). I had to wait to see if the toe turned black. Since it did not, the Kaslo hospital released me after a week. I returned home by bus to New Westminster – on crutches.

Before leaving to spend the summer recuperating up at Roberts Creek on the Sunshine Coast, I stopped for antibiotics at the drug store located at Eighth Avenue and Twentieth Street. I puked on to the floor right in front of the druggist. Because I had not earned enough to pay a full winter's tuition, I pumped gas while taking a few courses part-time. I kept working at the gas station the following summer (1954), learned to smoke and returned to UBC for my first year in the Bachelor of Science in Forestry program.

Chapter 3:

Summer 1955

In the summer of 1955, I went to work as a forestry assistant for the federal government in Manitoba. I took the train from Vancouver to Winnipeg, experienced the wind blowing at Portage and Main, and joined two other students plus the party chief. The latter looked forward to his wedding in the fall and hoped for a stress-free summer. We spent the summer putting in permanent sample plots and crawling over logged land in intense heat to measure the success of forest regeneration. The work area was the Sandilands Forest Reserve southeast of Winnipeg. Historically, the federal government established this forest parcel during the homesteading era to serve as a supply of building materials. After Manitoba assumed jurisdiction of federal Crown lands in 1930, Sandilands Forest Reserve stayed within federal jurisdiction, possibly to carry on as a forest research site. The tree species in place were jack pine, with red pine as an experimental introduction. During the first couple of months we lived in a farmhouse at Piney, complete with outhouse and Sears catalogue. Marchand, the community closest to Piney earned a reputation as the VD centre for southeast Manitoba. After that, we ended up living in the Hadashville forest nursery bunkhouse. We kept the butter cool in the White River. One Sunday, I went to retrieve some for our pancakes and reeled in the cord wrapped around the butter. I retrieved the package, but it was empty. That was a lesson in how not to keep butter cold. The River melted it and the contents disappeared downstream and out into the Lake of the Woods.

It was not my favorite summer. However, I learned to drive. Jim, the project leader, let me drive the Ford van while he hung from the rear door throwing out wooden stakes periodically across the experiment area. As is usual with new operators, I had trouble engaging the clutch. I recollect the occasion that Jim hurtled out of the back along with the bundles of stakes as I engaged the clutch. He still let me drive though.

That summer, I took a supplemental exam after flunking the surveying course back at UBC. I entered the classroom to write the exam at the University of Winnipeg. I had my briefcase under one arm. The official gave me the envelope containing the exam. Before leaving, he told me to leave the exam sheets on the desk after I finished. I knew I had my study notes in the briefcase, but they stayed there. I caught the bus back to Piney, and travelled through Steinbach. On the way up I had migraine. It evaporated on the way back. I got a passing grade.

I also learned to play bridge and drink beer. And the flat land was depressing. I got by pretending that the distant cloud banks were mountains. A tornado (June 20, 1955) blew apart the community of Vita. The crew travelled through the village shortly after. The hospital was without roof or windows, and mattresses rested at the tops of trees. One boy was killed and a little girl was rescued from part way up a boulevard tree.

One Sunday, midafternoon in August, a huge column of blackish smoke rose into the firmament to the south of us. Right away, we students were put to work on that forest fire. I was given to the forest ranger at Dawson Cabin. This was a ranger station located on the old Dawson Trail. It was part of the first All-Canadian route that linked up the Great Lakes to the Canadian prairie, and then on to Winnipeg. He put me to work as the dispatcher. Forest fire management was the responsibility of the province. I took orders from C.K. Smith, the legendary forester in charge of Manitoba forest fire control and T.R. Webster, the forester in charge of forest management in Manitoba. Early each morning I fried a stack of bacon to have it ready for C.K when he came in – my first task each morning – it had to be crispy. The forest ranger took me out with him to light backfires and invigorate local residents who were too lazy to fight fire. While I drove, he leaned out with his drip torch and ignited the bush across the road from their front doors. From then on, he gave me his truck to undertake inspections of the back roads to warn anyone that the fire was spreading north. It might even cut off the Trans-Canada. That was how I improved my driving skills. The truck (pickup) had a Manitoba Forest Service decal on each door. On one tour, I found a sugar daddy and his lady parked in an old gravel pit. I strutted up to his Cadillac and told him to get out of there pronto as a fire was coming up behind them. Disappointingly, they had their clothes on. They skedaddled back to Winnipeg, I suspect.

The northern edge of the Sandilands fire in August 1955 stopped just south of the town of Richer, on the south side of the Trans-Canada Highway. It was a subsistence community at the time. The feds had just booted most of the small forest operators out of the forest due to perceived overcutting. However, one little sawmill was still operating. Three men from Richer went back into their millsite to rescue the sawmill from the encroaching fire. They did not come home. Sometime later three charred bodies were found up against the sawdust pile at the millsite. The ranger had a photo on his desk as part of his investigation. One of them had a cast on his leg. The fire had left it brilliantly white in colour. We believed that the other two had stayed with him because of his disability. They could not start their truck.

Before I could leave for home, I had unfinished paperwork regarding a missing canteen to fill out. I do not know how it got lost. After I finished the explanation that it was run over by a caterpillar tractor, and I had thrown it away because it leaked, I completed the necessary form covering lost equipment. It was World War Two issue.

I was late returning to UBC that fall.

Ω

42

Chapter 4:

Summer 1956

In the summer of 1956, I worked for a consulting forester in Vancouver called Barney Johnson. My first assignment was at Monte Lake, between Armstrong and Kamloops. It was range country within the Nicola Valley plateau. The sawmill super, Tom Elliott, took the crew out to the worksite each day in his pickup. Most of them had to sit outside in the box. Tom chewed snoose. Sometimes he forgot to wind down his window, and at other times let out a wad of snoose that travelled back and connected with his passengers. The job was to remeasure permanent sample plots in Douglas fir stands to find out the change in growth rate since the last measurement five years earlier. I saw my first bobcat there. The job was over soon and Barney sent me to Quesnel.

We foresters can never look at a forest as just 'a forest'. Those growth rates tell us if we are looking at a high yield forest, a forest with medium growth, or one with low productivity. With today's ecosystem mapping, our job of upholding a sustainably managed forest gets easier.

After that assignment, I ended up on the Blackwater River west of Quesnel. I was the compassman and Albert Bush was the forest engineering tech. Our job was to flag in a logging road route for Patchett & Sons sawmill in Quesnel. Since it was early May, we camped in snow on top of an esker above the River. It was a hush-hush job in that the Forest Service had not yet put up that timber sale for bid. This was the forestry equivalent of claim jumping. Albert and I were dumped off at the work site and told not to return until we had completed the job. The boss provided us with food for a month. He instructed us to talk to no one who may show up. I developed 'squeak heel' before the completion of our work, a particularly painful inflammation of the Achilles tendon in my right ankle. Albert became frantic that we would not finish the job before his wedding day. In addition, we did indeed get a visit from a spy working for the competition. Albert was in a perpetual dither. He had left his mentor, Sig Techey, a forest engineer with one of the big coastal outfits so this was his first solo job. It was also my first experience with the intricacies around the use of a staff compass. During the last week we ran out of grub, so lived on spruce grouse, Cheez Whiz© and porridge. Albert vibrated day and night in anticipation of his big day. From there, I was dispatched to Prince George.

I must have arrived there by train: Paul Brett, party leader and new forestry graduate from UBC, Mike Angel, and I. Our job was to do a timber cruise of nineteen timber licenses on the McGregor River, otherwise known as the Cargill limits. Anyway, we all disembarked from the train via Prince George at the village of Hansard on the upper Fraser River. We brought all the gear and grub for an extended stay, including a spare outboard engine. Our contact was another person named Elliott. He ran the post office and hired out as a riverboat guide. He knew the McGregor like the back of his hand. Today, I can find Hansard on Google Earth.

After everything was loaded into the riverboat, Mr. Elliott swung downstream on the Fraser itself before getting to the junction with the McGregor. We travelled upstream to a proposed camp at the mouth of a large canyon up the McGregor. Mr. Elliott gave everyone a crash course on reading the River. We watched out for shallow corners and logjams. He also stressed that the River ran at nine mph even though the spring freshet was over. Once he unloaded us on a gravel bar at the canyon mouth he returned downstream with a commitment to resupply in ten days.

We stayed on the south side of the McGregor and completed the mapping and measuring involved in timber cruising. The objective was to return home with completed maps indicating the terrain and logging chance. We calculated timber volumes from the plot data collected. Then we grouped the plots according to timber volume and species. Afterwards, we drew polygon boundaries around each group of plots. Our timber licence map could now display the areas of 'good', 'medium' and 'poor 'timber volume categories. Then the area of each polygon provided the acres of each class of timber identified on the ground. We multiplied acres by the average volume per acre generated by the plot data for each polygon on the timber licence. The loggers could now take the timber volume estimates and plan their operations. Two of us worked at the job and the third stayed home to cook and do the field calculations from the measurements made the previous day. The timber was large spruce of good quality. But the Devil's Club[7] was pernicious.

Eventually we completed all the work downstream on the south side. Mr. Elliott arrived one day with a riverboat that he left at the camp on the gravel bar below the canyon. Thus, we could cross to the north side and complete the cruising of several more licences. We knew that at some stage someone would need to go upstream through the canyon to its top end to keep working up the north side. Mr. Elliott had warned us the River at the top of the canyon made a sharp right turn. The rough water indicated the presence of rocks just under the water surface. He told us to travel in the smooth strips of water beside the rough spots. We should stay in the middle once we arrived at the top end of the canyon, otherwise, we could hit the shallows and lose a shear pin.

I happened to be the first with the job of going up the canyon. I got little done that day as I fretted over that first excursion into the unknown. I dropped the boys off downstream that morning but they would finish their day's work above the canyon. I knew where exactly to pick them up since they showed me the rendezvous on the government air photos. At 4 pm, I got into the boat and proceeded up through the canyon. It was solid rock reaching up 25 metres on both sides; halfway up, the River made a right angle turn to the north. Mr. Elliott had informed us that if something happened in the canyon we would not crash into the canyon wall. The volume of water at that bend would cushion us and send us back into the mainstream. More than anything else, I was scared that the motor would not start or would quit part way up.

I did make it up through the canyon, felt the water at the dog's leg push me back from the cliff. However I did cut too close to the corner at the top end of the canyon, and broke the shear pin on the prop shaft when I had hit a submerged gravel bar. From there, I flew back down the canyon lying on my back looking up. The boat behaved like a piece of driftwood heading downstream, bobbing, spinning, and turning end for end. Shear pins are located between the propeller and the engine shaft. One can change a shear pin easily.

At the base of the canyon, the water was smooth once more. I got to shore, changed the pin and started out again. At least I now knew what the water was like up through the canyon. This time I got out of the canyon at the top end with no problem – until faced with a wall of water stretching right across the River, as in a weir. It must have been a submerged rock shelf. I just aimed right for it, got through perfectly, and was on time to pick up the boys. Don't forget that they had no knowledge of that wall of water. So coming home their startled looks when faced with this weir was my reward for the day. When faced with the turbulence in the canyon they spent the rest of the trip home huddled on the floorboards.

One day, Mike lost his cool during his everlasting battle with Devil's Club, insects and me. It was just too much, especially after he fell down a concealed bank once too often. He just cracked and threw away his compass. It hit a rock, the glass broke and the needle on its diamond pivot shot out into the brush. Mike and I had the air photo to help find a route back to the boat through miles of swamps and muskeg. On the way, I tripped and fell back on to a sharp branch stub sticking out of a windfall spruce. This was case hardened from so much time exposed to the weather. The stub skewered me right at the base of my tailbone. I had pain from that incident for years afterwards.

On another occasion, the three of us camped on a gravel bar at the mouth of a mountain creek. Camp spots were hard to find because the river banks were high and inaccessible. The water was deliciously cold. Paul put all the meat in the galvanized wash tub, partly immersed it using rocks, covered it with white butcher's cloth, and then tied it to the creek bank. We rolled two barrels of boat gas out on to the gravel bar, put up the tent and cooked supper in a downpour.

Part way through the night, we woke up to clanking sounds – bears? Oh no. A cloudburst up in the mountain brought down a wall of water and almost washed us and the camp into the river. We caught the mooring rope on the boat just as it was taking off. The fuel barrels were floating in an eddy just downstream along with the meat. Eventually, through wading and urgent poling, we recovered everything, even though it was dark. After that experience we learned never to camp close to the mouth of a creek again.

Then Barney's sidekick, Brooks Cranston, joined us. He brought along a professional Yukon River boatman. Barney was concerned that freeze up would start before the job was finished. This arrangement gave us two crews in the bush. We increased our daily traverses and timber measurements accordingly. The boatman was very uncomfortable on the river. He was scared stiff. He quit one morning right at the start of some rapids. He ended up staying in the camp for days until Mr. Elliott arrived on a routine supply trip and took him away. Brooks also brought up a case of sauerkraut. The tins were rusty and the labels had fallen off. We used the cans for target practice. Paul had tried to shoot a caribou as it was swimming across the river. Luckily he missed. Paul had brought with him the most professional of hunting rifles, a brand new Husqvarna .30-06. Brooks also brought up a rifle: a Ross .303 calibre World War One army rifle. The barrel had a bend in it.

By this time, it was late August or early September. We gained a cook. The time left for completing the work became critical as there was ice forming on the water. The poplars on the upper hillsides were turning yellow, and the mornings were frosty. Everyone practiced moose calls. We were camped on the north side of Herrick Creek. At the end of one day's work, Mike and I answered a moose call from the direction of camp. As we walked out into the opening on the creek below camp, we faced three would-be hunters aiming their rifles directly at us. No more moose calls for me after that.

Finally, Herrick Creek, the north fork of the north fork of the north fork of the McGregor River, dried up as the upper streams froze over. We portaged back down through the remaining pools until we reached the main river. Downstream we went, through the canyon, down and around the logjams until we reached Sinclair Mills on the

Fraser, just up from the McGregor River junction. After one night in the bunkhouse there, we got back to Prince George. I still remember packing that outboard motor over my shoulder as we walked away from the baggage car and the railway station in Prince George.

Before closing the account of my 1956 student summer, I have to mention that C.D Schultz and Co., a large forestry service company based in Vancouver, had also completed a timber cruising job further up the Fraser River. However, they chose to return to Prince George in the boats. Somewhere near the Giscome Rapids on the Fraser River, one of their boats overturned. A whole summer's data went floating downriver.

Once more, I was late returning to UBC for my third year in the Forestry Faculty.

Chapter 5:

Summer 1957, Canfor, Englewood logging division

An unavoidable technical note.

The language in this next section becomes technical by necessity. Remember, at the start of the 1957 summer, I was ignorant of how west coast logging worked. First, fell the timber and merchandise it into logs of varying lengths and top diameter by species. Second, drag them down to a central point (landing). Third, transfer the logs to a rail car or log truck for hauling down to the booming grounds. A summary of the technical aspects of highlead cable-yarding logging follows:

The Englewood logging process from stump to dump in 1957
(Reading this is optional.)

Englewood was a beach camp and log dump, on the north side of Beaver Cove, across from Alert Bay. Wood and English operated the log dump here before they sold out to the Bentleys and Prentices after 1938. These families created Canadian Forest Products Ltd. They extended the logging railway up the Nimpkish Valley. It ran eighty to ninety miles south to the headwaters of the Oktwanch River. The operation that spanned five camps became the families' Englewood logging division [8].

Canadian Forest Products Ltd (Canfor) Englewood logging division had one purpose: to supply logs to the Company's manufacturing plants. The plywood plant in New Westminster (foot of Braid Street) took delivery of log booms containing Douglas fir peelers. All the sawtimber, hemlock and Douglas fir, went to the Company's sawmill at Eburne, on the north arm of the Fraser River at the foot of Granville St., Vancouver. Three headrigs sawed logs sorted by size. The pulpmill at Port Mellon in Howe Sound received all the hemlock pulp log grades. Finally, the Hunting-Merritt cedar shingle division in New Westminster received all the western red cedar.

The BC Forest Service, through the Forest Act and regulations, refereed the harvesting process. The timber originated from gazetted Crown provincial forest lands; although Canfor added in its private timber licences to boost the regulated allowable cut of timber. The Crown relied on Crown timber dues collected to supply the government revenue stream. Crown-licensed timber scalers measured and graded all logs leaving the Nimpkish Tree Farm Licence. Then the Forest Service issued timber stumpage and royalty invoices (my job in 1951-52).

The production of logs
The BC Forest Service inventoried all Crown forestlands (as in my summer job in 1953). Note that foresters reconnoitre all applicable harvest areas before final application. They amend plans as necessary, especially where there is a potential for erosion or damage to fish-bearing streams. These applications also include culvert placement and size, and bridge specifications. Company timber cruisers revisit the proposed cutblocks and confirm timber volume by species and log grade. The engineers and forestry crews flag and paint the boundaries of each cutblock, ribbon in the access roads and mark the landings.

Next: The fallers fell timber on the rights-of-way. The construction department then pushes the down timber aside to build an access road. Gravel crews then spread gravel to provide a smooth hauling surface. Before this can happen, they install culverts and bridges where needed. Federal Fisheries have to approve all stream crossings. After that, mobile log loaders move in and transfer right-of-way timber on to log trucks for the journey out.

At some point, the forest technologists move in, paint the corner posts, and then ribbon in the logging boundaries. We used axe blazes since survey ribbon was yet to be invented.

Next, loggers move in. First, fallers and buckers whose task is to convert each tree to logs. They lay the timber down across the side hill 'in lead' with direction of pull chosen to prevent hang-ups and breakage when logs are yarded down to the landing. Then the highlead crews move in, set up the highlead towers and move in the log loader on the lowbed transporter.

The logging camp superintendent had the overall responsibility to move wood from the stump on to the log cars or logging trucks, thence to the reload. His staff (*):

*Camp timekeeper/administrator
*First aid man
*Master mechanic and crew
*Construction foreman and crew
*Bullbucker (falling and bucking foreman). In Vernon camp we had 50 fallers and scalers
*Woods foreman – he directed the logging and transportation
*Assistant woods foreman, sometimes
*Cookhouse staff: Cook baker, flunkies and bullcook

The woods foreman directed (-):

-Lowbed operator
-Loaders and operators
-A crew of loggers for each highlead set up
This included:
-Hooktender: in charge of a set of loggers. He also changed the placement of tailblocks that opened up a new yarding road. This man relied on the:
-Rigging slinger: who supervised the two chokermen attaching logs to the butt rigging in the logging area. He gave orders to the whistle punk.
-Whistle punk: his job was to transmit orders issued by the rigging slinger down to the yarding engineer at the landing by blowing a

whistle according to code. One whistle was 'stop', two whistles 'go ahead slow' and three whistles told the yarding engineer to apply power to the mainline drum, thus yarding the turn down to the landing. The punk relayed the signals to the landing with a long extension cord. He had a connection at his end that allowed him to activate the whistle at the landing. A steady set of short bursts on the whistle signaled an accident. All moving machinery and the butt rigging must stop immediately.

Then, at the landing:

-Yarding engineer

-loading engineer

-Head chaser (unhooked each turn of logs at the landing), and the second chaser

These latter men set the tongs on the heel boom to load the log on to a log car or logging truck. (Today, mobile loaders and log grapples.)

More terms

-Boomstick: 66 feet long with a minimum 8 inch top, containing the logs in a log raft

-Swifter: Same as a boomstick. These kept the boomsticks in a rectangle shape by acting as a brace across the log boom.

-Boom chains: heavy chains attached by a hole bored through at the end of each boomstick or swifter. A toggle at one end was slipped through the ring at the other end of the boomstick or swifter thus forming a log boom that could be towed without falling apart.

-A turn (of logs): The chokermen (two) attached chokers to logs in the bush, ready for yarding down to the landing. The chokers were attached to the butt rigging at the woods end of the mainline. They set anywhere between three and six chokers, or more, depending on the size of the timber.

-The rigup 'goat' was used with railway logging. Using this, a high rigger and rigup crew raised the spar tree, hung all the 'jewelry', threaded the mainline and haulback lines through the blocks at the top of the tree and prepared the landing for the commencement of logging. The goat was a flat deck rail car containing all the necessary straps, blocks, and pulleys. Also, it was equipped with a small donkey engine used to thread both the mainline and haulback lines through the sheaves strapped to the top of the spar tree. The crew used 'strawline' [9]for this function.

Figure 1 A boom of hemlock logs – photo source undetermined

Figure 2 A Cold-deck of logs waiting to be swung down to the landing. Source of photo unknown

Note the yarding machine on skids tucked in behind the spar tree.

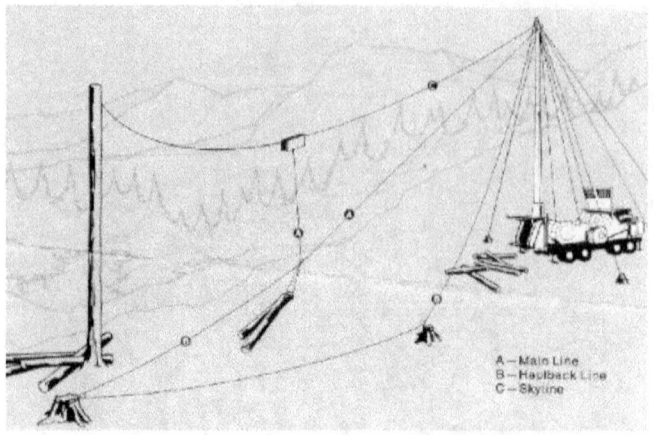

Figure 3 A typical North Bend yarding layout

Note the steel spar at the landing. This system, that includes a fixed skyline and carriage, can reach logs 1,400 feet out.

"One of the more innovative cable logging systems is known as the North Bend System. It is a suitable rigging system for a swinging cold-decker logs to the landing, but can also be used for logging as well. By anchoring the haulback line laterally to the skyline, the butt rigging can be pulled all over the place. The skyline is raised and stays in place. Consequently, though 3 lines are required, it can be rigged with a 2 drum machine, as the skyline can be raised and tied off, or anchored with a second machine. Practical working distance with this rigging method is 1400 feet with suitable deflection free of intervening high ground. It may be used on uphill, level or moderate downhill slopes. The load is divided between the skyline and the main line. Ordinarily, logs drag along the ground, but can be raised by holding the haulback line tight when an obstacle must be cleared." *(Source: http://www.vannattabros.com/cable3.html).* Permission requested for quote, but not replied to.

The rigup displayed had one advantage: it reached out further than the normal highlead setup.

Ω

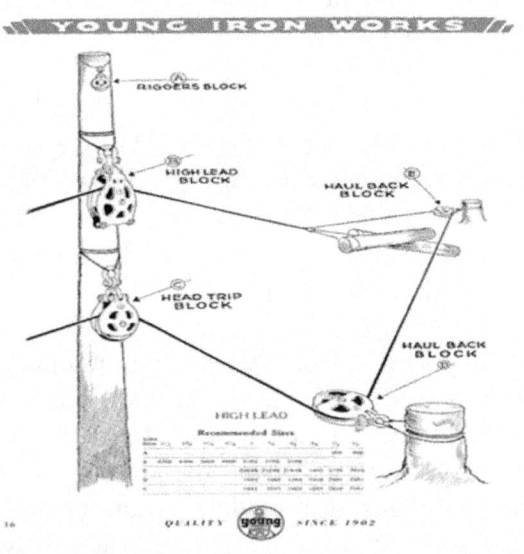

Figure 4 This set up is a conventional layout for hot yarding directly from the stump. Photo ex Young Iron Works catalogue

Summer 1957

In May 1957, Roy Jewesson hired me as summer student to study log breakage and waste. He was the divisional forester responsible for all forest management works on the Nimpkish Forest Management Licence. This covered the Englewood logging division of Canadian Forest Products Ltd. I flew to Port Hardy from Vancouver, and then rode in the Company Beaver aircraft (CF-IFQ) on floats piloted by Roy Berryman. We landed at the dock close to the divisional office at the south end of Nimpkish Lake south of Port McNeill.

The summer work program measured log breakage and waste from three separate logging systems. The logging methods sampled were: a) side 2's highlead yarding/loading operation at Woss camp that also included a cold-deck swing, b) cold-deck swing at Vernon camp using D8 caterpillar tractors equipped with logging arches, and c) a skyline skidder operation at camp A (highlead again, but with a difference).

A second student, Reiner Baum, joined me at Woss Camp. Reiner was a Teutonic Euro who rapidly became appalled at the volume of wasted and log breakage visible at the work site. Reiner and I started work at a highlead logging operation on H line up Hoomak Creek (spur HH). This meant that by 0715 hours we had eaten breakfast, made our lunches and, with 300 other souls, were aboard the speeders and 'crew cars' that would take us to work. The crew stowed its work tools in their allotted storage in racks bolted to the outside. Sometimes we travelled in a 'cattle car' pulled by a locie (locomotive). During fire season, the early shift was 0415 hours. Up to 11 men worked in this location, including us.

Our job was data collection. We split the job into two. The man back in the woods stayed with the chokermen and tallied the log species, top diameters and lengths in the 'turn' as the chokermen attached the chokers to the logs. To start the journey to the landing the rigging puller (slinger) yelled at the whistle punk to blow three hoots on the whistle. Following this, the engineer down below let off the brake. Then, the butt rigging rose majestically into the air, tight lining, clacking and banging into stumps as it fought its way to the landing. We captured measurements of length and top diameter as safety permitted. We estimated the rest. I had arrived in camp with brand new caulk boots (Daytons) to replace my Heads light cruisers. The man down at the landing (destination) tallied the logs and broken chunks that did arrive. We used scale sticks and calipers, together with 'eye balling' when not able to get close in. Boomsticks and swifters were always in short supply. They broke first because of their length. All four logging camps delivered their production to Beaver Cove for sorting and packaging into log booms for the journey to Vancouver

The fallers carefully felled and bucked trees with boomstick potential into 66-foot pieces. These were a specialty item in the array of log products produced. The call would go out from the Beach camp. The bullbucker and camp superintendent decided which cut areas (settings) contained the best boomstick material. As a tree severed itself across the sidehill, it fell on top of stumps or other fallen timber. Occasionally, this would create a bind just where the bucker had to make his boomstick cut. A bound tree could ricochet backwards and catch the bucker unaware. Both fallers and buckers lost their lives. They were paid on the volume scaled for each team.

This lead to 'bushelling' or felling the timber faster than was safe. The timber scaled after the bucking phase told the super how many boomsticks sat ready to yard into the landing. He relayed this information to the railway department, who then made up the night trains of empties returning to the landings. Most railcars were forty-footers, but a boomstick car was sixty-six feet long.

The logging equipment was all located on rail on a stub spur. A massive Douglas fir spar tree stood in place. Four, 1¼ -inch wire rope guy wires, wrapped around and spiked to the stumps using railroad spikes, kept it upright. On the inland side of this tree sat a yarding machine on a flat deck rail-car – tied down and fixed in place. Englewood had only lately switched to diesel from steam. The yarding engineer controlled a number of winches. He operated a Gearmatic winch for the mainline, another winch for the haulback line and a third for the strawline. A massive Cummins diesel engine supplied the power. A fourth winch holding the spotting line led off from the yarder to the empty log cars waiting for their turn under the loading boom.

The loading machine on a sled was strapped to the spar tree on the outbound side of the spar tree. Suspended to one side was a heelboom with two sets of tongs dangling from it. Beside the loading engineer, there was a head loader, a second loader and a third loader. Many logs, Douglas fir and cedar, were too large and had to be loaded on to a log car with straps, since the tongs were too small. Not to forget the chaser and second chaser who unhooked the turn of logs at the landing after its journey down the skidway.

Logging waste produced a loss in boomsticks, sawtimber piece counts and broken logs too short to merchandise. Also, waste occurred from logs and chunks that fell out of the chokers on the way down to the landing. After the finish of each yarding road, Reiner and I would scale up all the abandoned logs and chunks remaining on the skid road. Often a large cedar log would start its journey to the landing, only to be shattered against stumps on its way down. At the bottom, the yarding engineer delicately manoeuvred the turn of logs into the reach of the loader. The presence of log breakage was apparent as one choker arrived with no log attached. Cedar was worth $35 per one thousand board feet in those days, hardly worth it.

I remember my Hoomak occasion in several ways. As the sun crept up the mountain on an early shift morning, the ravens called to each as they spiraled up and up on the air currents rising from a warming valley. Then the echoes, oh, the echoes, contributed by the sounds from whistle signals followed up by the roar of the diesel engine at full throttle created echoes that bounced around and around the hillsides. The chokermen left pancakes on a stump for the bears except that, besides the honey, they added a healthy portion of pepper. Big cheers of encouragement arose as the bear went streaking across the side hill after its breakfast.

Eventually, we moved to a catside swing at Vernon camp – on the north end of the Vernon burn. Here the timber was mostly peeler grade Douglas fir. Four arch cats would skid the timber down from the upper decks. An arch-cat could break just as much or more timber, especially boomsticks. We worked in mud at times. Danger existed when a boomstick got bound up behind a stump, then whipped across the skid road as it came free. We refrained from following directly behind the turn coming downhill towards the landing. A fifth cat complete with cable blade did the bullcook work. All these machines were Caterpillar D 8s. The logging arch attached to each tractor kept the turn of logs suspended with a winchline, in order to facilitate the drag to the landing.

Figure 5 Arch cat swinging in a turn of logs. Source: www.mosaic.bc.com

Sometime after we had moved on to our next worksite, the landing crew was sitting one day on the brow log eating their lunches in the sun. They sat under the Mills boom loader. The brake toggle on the drum controlling the lift of the loading boom released. It fell and killed two of them.

Ω

In August, we were on the Camp A skidder up Kaipit Creek. The difference in yarding distance was significant with a skidder set up. First, there were back spars aligned like spokes of a wheel. The external distance to the back end was 1,800 feet. A fixed skyline, suspended between the home tree and back spar, supported the skyline carriage. The mainline ran up through a sheave on the carriage. It now swerved as a lift platform. The yarding engineer at the landing controlled the movement of the skyline carriage. He could haul it in with the chokers full of logs, or he could pull it back empty to the spot above the chokermen. By applying winch power to the haulback line, he dropped the butt rigging at their feet.

We had road access. We rode to work in an old crew bus. Log trucks trucked away the production as it was loaded out from the landing on its way to the Camp A reload. We had several storms that August. During one lightning storm a bolt of lightning passed up the whistle wire, and zapped the whistle punk. The bolt blew him into a nest of logs. We pulled him out unhurt. That same storm a little later appeared again as we were sitting in the bus on our way back to camp. We were all fast asleep in the bus with fogged up windows and full of loggers in wet clothing. The atmosphere was steaming inside that bus. Wham! Lightning struck a tree right beside us; a dazzling flash of bright blue light lit us up momentarily. The bus operator (yarding engineer making some OT) kept us on the road though, but no more shut-eye on that ride home. Everyone was yapping at everyone else all the way back to camp.

I spent the last of my summer in 1957 down at the forestry office organizing the data and consulting with Roy Jewesson. He was an exceptionally incredible boss. He was the exact opposite of a micro-manager, and I flourished in this environment. Working for Roy was a privilege and joy. He never intervened once we designed

59

the data sheets and I was given a free hand to conduct the investigation.

I recollect that it was a wet summer; although, the survey crew did have an escaped camp fire up KH line somewhere. I do not think that Reiner stayed in forestry. I believe his career was in the environmental arena – in the States?

Ω

I returned to UBC to my fourth year forestry thesis, centred on the analysis of the data collected that summer. There were no computers then, no up-to-date mechanical machines that added and subtracted and multiplied and divided. I did all the data crunching by hand. I analysed the data – oh for spreadsheets – and produced a report that Canadian Forest Products (Canfor) used to curtail cold-decking. It was an incredibly interesting summer. It taught me to stay alive – not sideswiped by logs that were in the bind and then let loose. We learned to stand behind the mainline instead of in front – stay out of the bite.

Interlude

However, before continuing with my forestry journey I pause to reflect on the next phase. This is perhaps better flagged as the "Hungarian" epoch. The entire Sopron Faculty of Forest Engineering based at the University of Sopron in Hungary fought its way to the Austrian border after the uprising against Russia in 1956. In the late winter of 1956-57 they arrived at an army camp in Chilliwack. There, they set up the Hungarian division of the Faculty of Forestry at UBC (ref: http://www.hungarianpresence.ca/History/Sopron – Kozak – pt4.cfm). They took the evening shift in our classrooms and labs. Our Dean George Allen and the federal minister, Hugh Pickersgill orchestrated the move to Canada.

We forestry students met the buses at the Chilliwack army camp and offered our welcome. However, we were there partly to size up the female portion of their student body – no lady foresters at UBC in 1957. The language barrier was significant at the start. It was awkward. I got up in the Dean's fourth year seminar on one occasion and complained bitterly the Hungarians were stealing the rare job offers available after graduation. We were the smallest forestry class since the big flow-through of World War II vets. They left our teaching faculty members shell-shocked and worn out. There were only 23 of us graduating in May 1958 compared with 150 each year previously. And there were only 3 job offers that year. Industry had stepped up to the plate and hired all the Sopron students as summer crews.

At the last minute, though, someone offered me a forestry fire warden's job at Englewood. I caught a Union Steamships passenger vessel full of loggers travelling to Beaver Cove, and thus my professional-in-training forestry life began. My boss left for Prince George shortly after and I took over his responsibilities temporarily. By this time (1961) we had six Sopron grads on our forestry staff at Englewood. The British Columbia forest industry, federal and provincial forestry agencies, university faculties across Canada and forest research all benefited from Sopron's entry into Canada. Later, the various institutes of technology (Prince George, Nanaimo and British Columbia Institute of Technology) took up the slack. With the Sopron men and women arriving just as the post-war lumber market exploded, Canada and British Columbia would benefit greatly from the astute move made by the team of Dean George Allen and Honourable Hugh Pickersgill.

Ω

Chapter 6:

Return to the Nimpkish Valley, 1958 (Canfor, Englewood logging division)

The Nimpkish Valley,
Vancouver Island. B.C.

Greetings

Figure 6 Mount Cain, 1965 or so. Photo by Eric Cooke, lathe operator, Nimpkish car shop.

Glen Patterson had offered me that fire warden job, and I took it. By this time, my bank account had a balance of $2.25. After I had given the steward from the Union Steamship 25 cents on the dock at Beaver Cove, my bank account shrunk some more. Remember Canadian Forest Products speeder 121? It was the Greyhound bus of the railway. Martin Headman and Johnny Danshin were two of the senior operators. We all clambered aboard and began our bone-jolting trip to Nimpkish, and then on to the destination at Woss Camp. Does anyone remember the bench seat that ran the length of the speeder in the middle? The angle iron bolted on to the edges of the bench? That iron strip was the bottle opener for all the beer consumed on the way.

Passengers, who could not remove the beer cap, merely broke off the top on the iron strip – or used their teeth – then poured the contents down the hatch anyway.

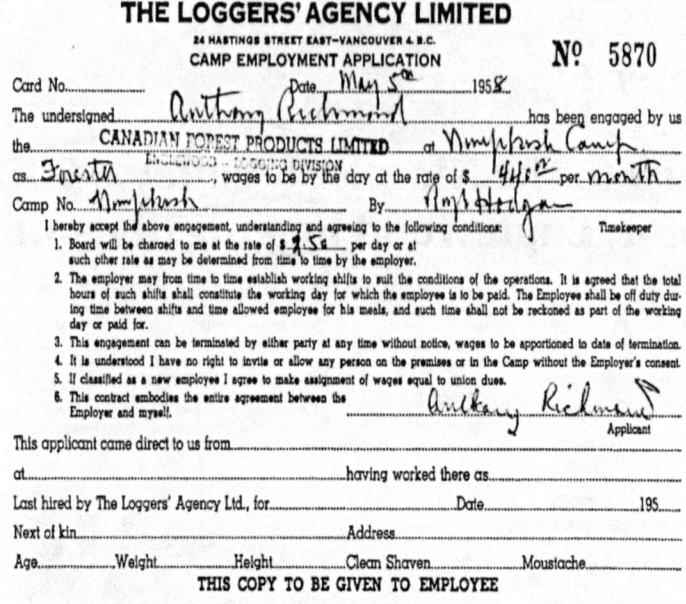

Figure 7 TR's hiring chit, 1958

At the marshalling yard in Woss, the speeder disgorged its load of semi horizontal people. They then crawled or walked over to their respective reserved beds and nests in the bunkhouses. The fallers, treated like management, had their own bunkhouse locations – one guy to a room.

Surveyors inhabited their own preferred lodgings. Seniority was in play, as was rank. Timber scalers, forestry technicians, machine operators took up the space left over. Senior employees got their own room. My destiny, as a newbie, was a room shared with a Russian faller who snored loudly and continuously, and who wore woollen Stanfield long johns year round. We were close to the cookhouse, so I did have that perk at least.

The Englewood logging division built everything on skids. I do mean every building owned by Canfor, including even the married quarters, main office in Nimpkish, and the train dispatcher's office at Woss, the fire shop and whatever. The reason: as long as buildings were on skids they were classed as portable and attracted no property taxes. We all had 15 minutes to eat breakfast and find our transport to work by 0715 hours. The powersaw shop stood beside the crew marshalling area. The warehouse was also close by.

The camp relied on steam heat delivered through large, insulated pipes. The steam came from the steam plant housing a boiler fueled with Bunker C. This, in turn, arrived by tanker car from Beaver Cove and sat in a siding beside the plant. Nearby, the power plant housing the large diesel generator kept us in electricity year-round. Dunc Forbes was the camp super.

I lived in Woss camp because of the location of the fire shop. However, my turf extended from Beaver Cove, Beach camp, up to the end of the rail beyond Vernon camp – some 90 miles.

Woss fire shop

When I got to Woss, Glen Patterson's[10] aura still occupied the atmosphere. Glen was Englewood's first forester and he was renowned for travelling everywhere in his running shoes. Nothing was ever too irrelevant to examine and inspect thoroughly. Folks took note to determine if I also exuded the same chemistry. I disguised myself with my Dayton caulks, stagged off pants and braces. My hard hat was from the fire shop and white with the label 'Tony' in the front with a black felt marker. Oh yes; add the gloves tucked into the rear pocket of my jeans, stagged-off above the ankle. No belt, just suspenders.

The fire shop had glass panels in the roof. There was a small office part. Most of the shop, perhaps up to 80 feet long in two sections, contained 150,000 feet of 1½" hose. It also warehoused another 100,000 feet of 1 inch, some 30 Pacific Y backpack pumps (1½" diam.). It contained shelves full of Pacific 5A pumps (1 inch). It housed all the hand tools legally required on all logging operations, and then some. It maintained Ansul fire extinguishers filled with dry chemical, 10 pound and 20 pound for use on class A fires, and add in a supply of drip torches for slash burning. The fire extinguishers that contained carbon tetrachloride in the shops, offices, the married quarters within all five logging camps and on the logging equipment throughout the division were replaced with the dry chemical Ansuls. The rail spur on one side of us, accessed by a loading dock, held two 8,000-gallon tankers with the 500 gallon per minute pumps attached. I do not remember if these tanker cars filled with water served as sprinkler cars attached to log trains in dangerous fire weather. Brake shoe fires were unavoidable on long grades, on early shift afternoons.

We were the central fire equipment supply depot for four logging camps in the valley, and Beach camp. Our job had three components: a) the organization and provision of mandatory fire equipment to every logging operation and woods machine in the whole division; b) responsibility for structural fire protection throughout, which included the married quarters, and c) responsibility for the slash burning organization/administration/implementation arm within the Englewood division (Forest Act, Section 17). We were also responsible for two lookouts found in the Valley: the first called Gatepost and the second was the lookout on a big ridge north of Vernon camp between the Nimpkish River and Vernon Lake.

Woss Fire Protection was responsible for:

Warehousing all fire protection equipment in the division. This included the pumps and hose mentioned above, all hand tools, piss-cans, fire extinguishers, the tank cars, and eight fire trucks. The railway also had sprinkler cars that ran behind the log trains on steeper grades. As well, Woss Fire Protection was responsible for ensuring that all fire season tools and equipment were in place between Beaver Cove and Vernon camp. The one responsibility that was mine alone was for the successful result of all Forest Service inspections covering fire, tools, and logging. Forget Easter; it was one of the busiest times of the year for Cec and me. We repaired or installed all the regulatory forest fire equipment on anything in the Valley that could throw sparks. Everything: Chainsaws, cats, power shovels, locies, crew vehicles, construction equipment, the few cold-deckers. We serviced all water barrels on railway bridges, not to forget buckets, and so forth. The Regulations dictated this had to be in place by May 1, the official start of fire season. Knowing that Mother Nature goes by her own rules, we attempted to have all critical equipment in place by Easter.

Once fire season was over, and all equipment cleaned up and put back in storage, we concentrated on fire protection in all structures. These included the married quarters from one end of the claim to the other. Not to forget fire trucks in each camp, and fire hydrants. I remember attending a course put on by the Vancouver Fire Department that added to my knowledge of fire protection for structures. Add in weather station maintenance.

We also disposed of 1,000 ha of logging slash each year through broadcast burning.

Cecil Frampton

Cec was my fire protection mentor and guru. I suspect Glen Patterson hired him, but this is just a guess. Cec originated by way of ranger duties with the Forest Service at Echo Bay, a boat district. He started with the Forest Service in the early forties.

Cec was a lifelong bachelor until he met and married Diane late in life. I was his best man. This occurred after he retired to White Rock. He had moved out of his little house on the Nimpkish River side of Woss camp. The camp weather station was not far away and either he or I took the three daily weather readings. We measured precipitation, wind, R/H[11], fuel moisture sticks (in fire season)[12], air temperatures plus maxi/min daily. The station conformed to government standards and was contained in a Stevenson screen surrounded by a high wire mesh fence.

Anyway, Cec taught me all aspects of fire behaviour and danger. We burned slash each fall using whatever crews we were given. My best results were with Louis Dempsey's cold-deck crew out of Woss. Also, Cec was inquisitive. We collected data to understand the correlation of soil moisture to hazard stick moisture levels. The purpose? To produce a more sensitive fire danger forecast. I became Cecil's protégé. He got me teaching various subjects at the annual forest fire control courses held at the Pacific National Exhibition, Vancouver, in the winter. I have a plaque from the 1960 course (5th year). The Coast Forest Fire Protection committee issued The 'Fire Control Notes' each year following each course. I wrote some of these.

Woss bridge fire.

1958 was a severe fire season year. The low relative humidifies created brake shoe fires on the railway, sleeper lightning fires all over, and the Woss bridge fire that destroyed it.

It was Sunday, May 4, just after 4 pm, when I got word the bridge deck was on fire. This bridge, crossing the Nimpkish River, at the south end of the Woss yard, was up against the division's diesel storage tanks. A spur line provided the access for the three diesel electric locomotives to refuel. Spillage had occurred over the years as tanks overflowed frequently. The ground at the site was saturated with diesel. As well, the underlying rock shelf directed a steady seeping of diesel over the bank into the Nimpkish River, some 50 feet below. This was the site of a great steelhead fishery just downstream.

Russell Mills, who was still the division manager when I first got there, designed the bridge. His bridge specs included huge Douglas fir stringers flattened on two sides with adzes. A bridge shop at Nimpkish built all stringers and skids used throughout the division. An underpinning of poles and piling supported the bridge deck. Although it was a railway bridge, added planking expanded its use for vehicle and foot traffic.

At 4 pm on this early Sunday in May, a dropped cigarette ignited a fire in the wood fuzz built up over the years between the rail and planking. The early efforts to put the fire out ran into trouble: the hose stream was not reaching the fire, and the fire was running under the bridge deck and was not reachable with water.

I ran to the fire shop and returned with a Pacific Y pump, gas and a suction hose. Folks lowered me down over the cliff, followed by hose, pump and suction. I then set the pump up on the riverbank, got it started, but no suction. The gasket must have fallen out en route. Therefore, I climbed back up, ran to the fire shop and returned with a pocketful of suction gaskets. I got the nozzle stream to bear on the underside of the bridge, but no use. The fire was spreading well out over the middle of the bridge span over the River and I could not reach it. I had to use fire hoses to scale the cliff, both up and down. That was my first fire in the Nimpkish – a total failure. I remember all this as if it was yesterday. Not one pump left the fire shop after that without a suction gasket tied to it.

Howard Elder, forest engineer in the division, designed the new bridge. I have a Jack Pacey print of two of the replacement bridge stringers ready to leave the bridge shop in Nimpkish.

Figure 8 Bridge shop, Nimpkish, 1959: two log stringers loaded and en route to replacement bridge site, Woss. Jack Pacey photo found in a garbage can.

Owen Hennigar's lightning fire at Camp A

Owen Hennigar was the Camp A superintendent. He had a fire up on a bluff just off one of his access roads. My job was to deliver 5A pumps, hose, and gas and relay tanks. It took two relay tank installations before we could get water to the fire. This occasion became one of my key learning experiences. Until then, I had no knowledge of the hydraulics attached to lifting water. A ½ pound of pressure = one foot of lift. I have thanked Owen for that lesson all the rest of my life. Tied into that is the subject of centrifugal pumping versus positive displacement pumping. I had a great deal to learn. I remain a strong proponent of Pacific Y pumps; although these are no longer manufactured. Their place is now usurped by Wajax© centrifugal pumps of assorted sizes.

The summer of 1958 was a scorcher. We went 41 days straight without rain. One can understand the actions of the hooktender who

69

pissed on the fuel moisture sticks to prevent them from getting down to a shutdown reading (10% fuel moisture). One summer each twelve years brings excessively dry weather for an extended time. A high-pressure air mass over the Central Interior blocks moister air approaching from the Pacific Ocean. Dry air blows down the coastal fjords and inlets, and penetrates the large coastal terrain. The Nimpkish Valley becomes a funnel for hot air flowing west down Knights Inlet. Daytime temperatures soar, intense drying follows, and easterly winds make it worse. Loggers cannot work after the relative humidity falls to thirty percent. The risk of fires from sparks or cigarettes is too high. First, they go on early shift to miss the hottest part of the day. Then, as that omega high-pressure air mass block continues, camps shutdown and everyone goes home.

A side effect of this moisture pattern is that soil moisture is depleted, and tree species go into stress mode. The outcome shows up the following spring. Then the stressed out forests produce enormous quantities of seed – Nature's way of sustaining the biomass.

Meeting Earl Laird for the first time.

I first met Trainmaster Earl Laird shortly after taking up my duties. The railway department was not immune to the legislated regulations regarding fire season. Spark arrestors, fire extinguishers, hand tools and piss-cans were part of fire season equipment for all locies and speeders (gasoline powered railway vehicle). The three diesel electric locies converted one fuel tank to hold emergency water during fire season.

One morning I finally found the yard engine (locie no. 112) used to make up trains and move cars to the warehouse, steam plant and cookhouse, parked at the locie shop at the south end of the yard.

Figure 9 Locie 112. Switch engine Woss yard. Englewood railway website.

I had just finished loading up the front platform with its designated fire season equipment, when up walks a guy my height wearing a clean white hardhat, ironed shirt with pens sticking out of his shirt pocket and clean gloves tucked into the back pocket of jeans held up by both a belt and suspenders. The two-storey dispatch shack (on skids) was in clear view so the dispatcher had probably alerted Earl whose office was downstairs. I got a huge chewing out delivered concisely and to the point. He then took me down into the pit and showed me the rack under the boiler, built to hold the fire equipment. I can see Earl right now – no unnecessary meat on his frame, a large nose, alert eyes and chiseled face. He soon was a key component in my ability to come and go to all the places between Beaver Cove and Sutton reload – a distance of 90 miles. I needed access to rail transport to get my job done. I had not realised this at that moment.

Ω

I do not know why exactly, but that night I wrote to Dean George Allen and the Faculty of Forestry at UBC to thank them for their efforts on our behalf during the last four years. After my contact moment with Earl I realised that 90 percent of my job depended on others. Earl was a 'keeper'.

Figure 10 Alec Ingertilla at home in the bunkhouse.

Alec Ingertilla

I do not know how this came about; just behind the office was a bunkhouse with just two rooms. It was equipped with a toilet and shower. I can only surmise that the chief dispatcher, Alec Ingertilla had elected me to take the other room and pulled the strings to make this happen. It was excellent accommodation for single workers. I really needed that 'in' with the railway to get to all locations in the valley. Truck roads were just beginning to augment rail.

Alec was a Finn who belonged to Alcoholics Anonymous. He turned himself out neat and clean. He lived for his women during shut downs and had a Polaroid camera to capture his current heart's desire on film. He was a big overweight man. He taught me how difficult the struggle to stay sober was.

Having this connection to the railway provided me with all the transport destinations I needed. This included the loan of my own little speeder. I travelled everywhere: The dispatcher would dispatch me to a certain junction – I might have to stay there until a log train of empties went by. I phoned the dispatcher and got my clearance to the next junction, and so I progressed to my destination. Also, I hitched rides on locies and speeders as needed and manned the switches for the trains.

For now, this concludes 1958. The dispatchers treated Roy Jewesson poorly. This bothered me off and on. They left him stranded at times, as if to make point that his role in the Valley was insignificant compared to theirs. They were totally unaware that without his role, there would not have been logs to load on to those rail cars.

Ω

My key role during fire season was to ensure that all forest fire regulations within the Forest Act were followed. This meant travelling to, then inspecting, cajoling and begging every industrial operation in the Nimpkish Valley to observe compliance. Management was always under threat of having some zealous forest officer shut down a logging operation due to noncompliance. Since fire season officially lasted from May 1 to September 30, I could not relax my vigil.

Ω

Chapter 7:

Nimpkish Valley, 1959-60

Figure 9 Woss camp Safety Week, `1959, Jack Pacey photo

Left to right: 2 bystanders, me, 2 fallers, Jack Vettleson (woods foreman), Gus Bavis (division engineer), Earl Laird (Trainmaster), 2 fallers, first aid man, Dunc Forbes (superintendent), school teacher, kids, crew.

Figure 10 Woss camp Side 1 set up at KH spur. Highlead yarding. Jack Pacey photo.

1959

Most of 1959 remains as a background 'hum' in my consciousness. I'll introduce Buck Dunsmore here, assistant forest ranger based in Alert Bay. Perhaps, I might be bringing him in too early though. I phoned Glen Bertram today since I could not remember Buck's name. It has been 50-odd years. Glen was the forest ranger in charge of the Northern Vancouver Island district and his office was in Alert Bay. Anyway, Buck, a true highlander and ex British Army I would think, was the heart and soul of 'rule by the book'. He was a fanatic when he applied all the regulations on fire tools and logging (snags, abandoned logs, dirty landings, and you name it.). We return to Buck in 1961. Meanwhile, I remember crawling up a sidehill at Buck's insistence; I think it was Mount Cain yellow cedar, to inspect this large yellow cedar log. From the road below, it resembled an abandoned log – lost revenue to the Crown. You can guess the rest – a gun barrel, hollow, and a true cull. The unwritten rule of course was to keep Buck from pulling rank and shutting down a camp or side due to negligent logging. Also, I recollect Buck going on a binge of snag slaying. He loathed seeing anything standing after logging. I had to consider the bullbucker's reluctance to pay for cull timber. Somehow, we got through these episodes. He would always re-inspect the cutovers to ensure that we had carried out his instructions. Of equal importance was the need to cool down the bullbucker and get him to buy into Buck Dunsmore's rulings. Safety rules dictated that no snag was to be left standing. Otherwise, it could be knocked over by the butt rigging and kill a chokerman. I ended up in a negotiator/facilitator role most of the time.

We were unaware that it was Roy's last full year in the Nimpkish. He spent much time rewriting the Forest Management Agreement for change-over into the Nimpkish Tree Farm Licence. It was partly a game of semantics. Hubert Bunce was still silviculture forester. This meant he managed the work of establishing the replacement forests. This involved growth, composition, health, and quality of forests to meet diverse needs and values. The division was going full bore on the rail-to-truck conversion process. The Hungarian contingent (Sopron) was fully occupied with cruising timber and forest regeneration surveys.

One memory of Roy was when he joined the Louis Dempsey cold-deck crew and me slashburning a very sensitive setting right beside Bridge 15, crossing the Davie River, a railway span on wooden piling. We had to light it up slowly and allow the fire to burn out its fireguard beside the bridge. I am sure we had a tanker car standing by. Anyway, we were well into the light up. I placed Roy in the line as we moved backwards from pile to pile across the setting. It was burning well, and then I noticed a gap appearing between the slashburners. No Roy. There he was in front of the fire front with a sling hygrometer measuring the ambient relative humidity. It was zero. After a while, he got bored and got a ride back to camp. However, this was one of the great fire behaviour lessons of my life. Spatial burning influences local climate mightily. I reference this phenomenon further in my account of the Muchalat fire (1965).

Ray Savola and his wife Phyllis allowed me to hang around their place in the Woss married quarters – but I confess to not knowing Ray's responsibilities in the Nimpkish forestry hierarchy.

1960
Our silviculture forester, Hubert Bunce, left to get married at some point, and then returned with his new wife on a sailboat from

77

Vancouver. He triggered great panic when he was overdue, but I cannot leave him to tell that tale since he passed away recently. I do remember spending a weekend with him and Gill sailing up and down the inlets behind Alert Bay. Hubert left to attend Syracuse University to pursue grad studies in forestry. Gill told me recently that they had fought through storms on their way up through the Salish Sea and had to anchor at times to avoid the weather.

So, I started to pick up the silviculture work where Hubert had left off. My first plantation was at Sutton Creek, at the south end of the 1953 Vernon burn. The fire had depleted all duff and humus, leaving bare rock and mineral soil. It burned so hot that outcrops of granite exfoliated. With its west-facing slope, the aspect was severe, even for Douglas fir. The tree plant foreman and railway section men were the tree planters. All Italians, except for the foreman – he was from Trieste. The seedlings came via deck cargo to Beaver Cove, then by flatcar to Sutton Creek. The trees arrived in bundles, wrapped in burlap with the seedling roots in the centre, and held together with wooden stickers and metal strapping locking everything firmly in place. It was an excellent method for transporting 2-0 bare rootstock.[13]

At Sutton Creek (spur KY), we threw the bundles into the creek to thaw them out. After that, we heeled them in along the edges of cat roads next to the landing. It had been a catside logging operation (logging was with arch cats rather than highlead). The planting mattocks were resurfaced and repaired before planting started about March 15 – my guess. I insisted that a small rock be placed at each tree to provide shade at the root collar. We planted at 10' by 10' spacing. It was important to keep the tree roots moistened because the planting strips were long and the only water was in Sutton Creek. I have thought for years that my ashes would be spread here eventually. However, with no more Canfor presence in the Valley, my chosen location is now at the top of Mount Elphinstone here on the Sunshine Coast.

Cruising for Plus trees and scion collection
There was a big push to improve seedling quality for Douglas fir on the West Coast. I learned to talk 'genetics' and recognise 'phenotypes'[14]. The Tree Improvement Committee – a group of

78

foresters representing all the big outfits set up Plus Tree cruising in Douglas fir timber stands aged from 100 to 140 years. This meant that we had to traverse each forest and inspect each tree for superior qualities – size, height, degree of branchiness and other sexy attributes. We then measured the chosen candidates and bored them to find out age, wood quality and health. Following this determination, a forester returned to each candidate and shot off scion material from the tree crown. The scions found themselves in a tree orchard grafted on to more humble root stock.

The rationale for all this was that female flowers would be pollinated by a mix of superior trees, thus producing Douglas fir tree seeds of superior growth characteristics. The Forest Service allowed companies an increase in harvest volume schedule (allowable cut) based on the probability that fibre yield per ha would jump by some negotiated percentage in new plantations spawned from this superior tree seed.

The first Douglas fir seed orchard in Canada was placed at the BC Forest Service research station at Cowichan on Vancouver Island. Dr. Alan Orr-Ewing was the forest geneticist leading the project. Now, Douglas fir commences to flower at around age twelve. But the Cowichan seed orchard contained a few trees of a promiscuous nature. These had flowers by age six or so. Thus Allan Orr-Ewing started to produce superior seed in half the time allotted.

The companies, including Canfor, started to build their seed orchards using the scion material collected. In my case, I found myself in March high above Powell River and wading around in four feet of wet snow. I had ordered 5,000 rounds of .22 long-rifle ammunition from head office. A telex reply asked if I was preparing to start a war somewhere close by. Those high elevation Douglas fir trees in upper elevations had different flowering habits from low elevation phenotypes. Thus, it was important not to mix that scion material from two distinctive elevations.

I struggled up through that snowpack, chose five superior trees, shot the scion material off the tree tops, returned to the plane, and was back on Nimpkish Lake in time for supper. That scion material ended in the Canfor clone bank on rail spur KH.

79

Back to the tree planting project at Sutton Creek. A few years ago, I came across that Council of Forest Industries book that pictured the progress of reforestation after logging. There was a picture of that sidehill taken from across the Valley. The landscape had healed, with my plantation looking very advanced and healthy. That is a location I would love to visit. I fear that it will be logged before my return.

This was the start of a truly dynamic time for me. Hubert had left, then that summer Roy started going up to Prince George. Over time he assembled the timber base for two brand new Canfor pulpmills and seven sawmills. The latter also served as the woodchip supply to these new entities. Then there was the administration of the Nimpkish Tree Farm Licence to maintain.

But, Russell Mills, the general manager based in Nimpkish did get me to do a 10-year wood supply scenario for the Englewood logging division. Mr. Mills was a legend; a graduate of the U. of Washington in Seattle. He had had a lifetime of engineering those great bridges associated with railway logging shows, and he designed and built all those bridges along K line[15]. Mr. Mills was colour blind, which was unknown to me. He wore glasses with yellow-tinted lenses, so I should have known. I located the next 10 years' worth of timber far up the creeks and side valleys to promote fire protection access. I separated out each year's cut area using pencil crayons. My subliminal purpose was the need to open up the Valley for forest fire initial attack purposes. What an amateur I turned out to be. The plan added an extra one hundred odd miles to the road building budget. Again, this whole exercise was a thrust to the learning curve.

At some time in that time period – Hubert had left and Roy was in Prince George – Head Office fired Russell Mills. Henry Hansen, the falling and bucking superintendent, took his place. This may have happened in early 1961 – I do not remember. All I do remember was the awful shock I got when Mr. Mills left. I was nowhere near Nimpkish when this happened, probably still based in Woss, but a huge shock to the system. From that moment on, I have adopted the policy of never considering that any job in the industry is a sinecure. The folks with the big jobs got spit out at a moment's notice.

Our forestry crew was, beside me, Cecil, Jimmy Ross our fire technician (mainly because his hardhat was black), and the Sopron boys: Steve Tolnai, Zoltan Fulopp, Joe Bako, Julius Kapitany and Tibor Kocsis. I was so fortunate to have such a competent crew. They were the backbone to everything 'forestry' in the Valley:

- Timber cruising
- Fire protection/slash burning
- Tree improvement/plus tree surveys all over the South coast in cooperation with other companies
- Establishment of the Nimpkish tree seed orchard and clone bank
- Tree planting
- Regeneration surveys
- Right-of-way spraying for brush control, both truck road and rail
- Spot seeding trials
- Nimpkish thinning plots
- Deer population surveys; revisiting fixed plots yearly and counting deer shit
- Photographic analyses to obtain log boom inventory data (Roy developed this system)
- Forest fire patrols by air
- Contractor admin (Dan Hanuse at Englewood)
- Forest nursery liaison
- Cone picking
- Plantation survival surveys, assessments and inventory
- Deer population research
- Research plots – maintenance and annual measurements
- Joe Bako did the seed orchard grafting on K H
- Ed Mulock initiated forest regeneration mapping and inventory.

Ω

I need to mention Jimmy once again. It is a mystery how he came to the Nimpkish and found himself attached to the fire shop. He was a little guy, spoke in the broadest of Scottish accents, and devoted his off hours to fishing using the most expensive rod and reel. He was a loner and dressed in khaki pants and shirt. I guess that he had originated in the Outer Hebrides. He was ageless.

Sometime around this time, I was wined and dined by Gerry Burch, the forestry vp for British Columbia Forest Products Ltd as a prelude to joining his world. I regretfully declined.

Ω

Chapter 8:

Nimpkish Valley, 1961-64

1961

Mr. Mills' replacement, Henry Hansen, was in play, at least by fire season. This left me and the Sopron boys to carry out all forest management duties connected to the Nimpkish Tree Farm Licence. I acted as the administration forester. We had another dry summer, with a nasty lightning storm that passed down the Nimpkish River system during summer shut down. There was just nobody around to use as initial attack. I remember Art Newman and a crew swathing their way up the Tsitika River to attack one fire. I remember three weeks straight on fire patrol with Roy Berryman and the Beaver aircraft CF-IFQ, equipped with floats – what a workhorse. We flew at 4,000 feet. Henry had told me to quit but I prevailed on him to allow me up one more time, and we spotted a smoke plume way up in the ridge west of Vernon Lake. Ole Jernberg, the Vernon camp superintendent, corralled a Bell bubble helicopter and flew pumps and hose to the ridge top above the fire. The rest of us clambered up the sidehill to the top and located the fire. Lightning had hit a yellow cedar and was burning deep into the duff. Assistant forest ranger, Buck Dunsmore from Alert Bay was with us and he dug a personal fire line that was at least 4 feet deep. I will never forget it. The fire was easy to put out – we gravity fed the water down and drowned the fire. However, Buck's reputation as a fireline digger remains to this day. Glen Bertram tells me that Buck went back to Scotland eventually.

Ω

Figure 11 June 1961. Two scouts and I were overdue from an overnight hike to earn badges.

One summer day, I accompanied two of my Scout troop on a weekend hike to earn badges. On Saturday morning we drove to Schoen Lake, then planned to hike back in the timber on the south side of the Davie River. We spent two nights camping out instead of one. The road was on the north side. We had to walk until we could find a bridge crossing. The whole claim was looking for us by daylight, Monday morning. Just two pairs of young legs that powered out before we could get back home.

Ω

On another occasion with Roy Jewesson, we took the forestry boat down to Camp L, and then walked down to the old Schilling homestead at the mouth of the Nimpkish River. We had to burn down the house to prevent squatters moving in and possibly setting fires that could escape into the bordering forest. It was a great pity to erase history; the house was two storey, and cladded with no.1 cedar tongue and groove siding. Roy piled up some cardboard under the stairwell. In four minutes the smoke started to erupt from under the eaves. We also threw stuff down the well as a safety precaution. It was a sad task. We walked home along the abandoned mainline from Englewood to old camp L. On the way, we passed a primitive camp of hippies – a mum, dad and a little toddler. The latter had no clothes on, and mosquito bites covered his body.

Luckily, the Lake was quiet during the ride back. However, it was still a twelve-hour day.

My Alert Bay Indian connections.

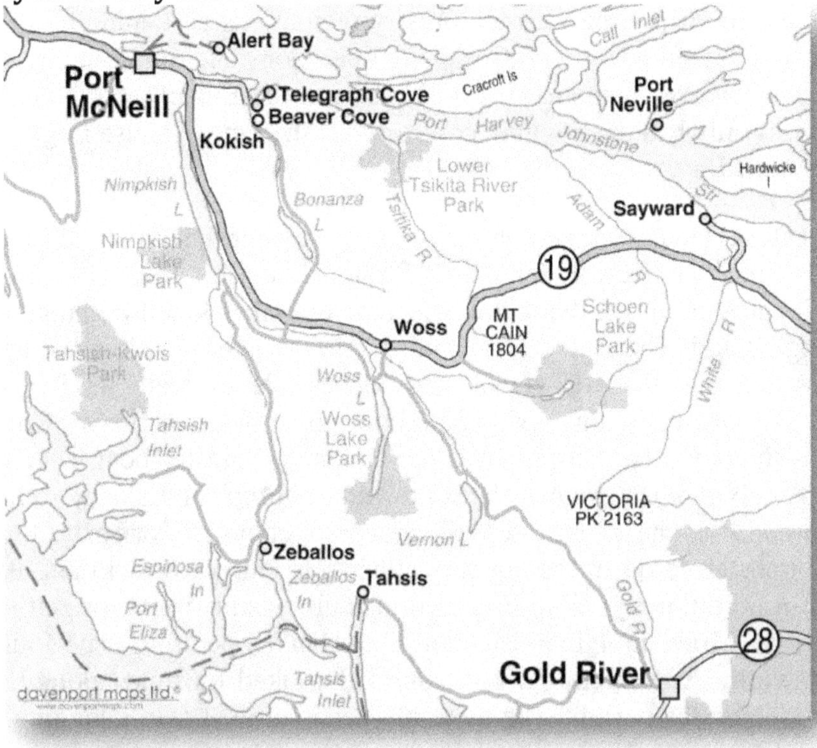

Figure 12 Nimpkish Valley from Beaver Cove south beyond Vernon Lake south to Gold River. Map© contributed by Davenport Maps, Victoria, BC.

Alert Bay has a place in my heart for several reasons. My daughter was born there, my first Indian, Dan Hanuse based his logging company there, and I got to meet Terry Dick, a rigging slinger, during my logging superintendent days. He also came from Alert Bay. Then there was Glen Bertram, district forest ranger, whose office and Forest Service boat were also located there.

Dan Hanuse.

During my early career in the Nimpkish Valley – it started in 1957 and ended in 1967 – I went from forest fire protection forester to silviculture forester growing replacement crops of trees following logging. This was followed up by a time as administration forester for Nimpkish Tree Farm Licence 37 covering 1 million acres. This yielded four million cubic metres of conifer logs annually. Thus I was part of a large scale forestry operation on the north end of Vancouver

Island that ignored its place within Indian traditional landscapes. My duty was to ensure the various forestry regulations pertaining to TFL 37 were adhered to by both Company operations and our contractors. However, we (Canadian Forest Products Ltd, Englewood logging division) only had one logging contractor at the time, Hanuse Logging from Alert Bay.

Dan Hanuse logged in an isolated corner known as Mount Holdsworth, just across Johnstone Strait from Alert Bay. His camp was located in an old bunkhouse in the semi-abandoned community of Englewood. The area was a thriving logging venue back in the late 1930s and into the 40s. It was only accessible by water taxi from Alert Bay or by hitching a ride on a boom boat based at Beaver Cove across the bay. My transport was the Company forestry boat. I drove this from Nimpkish camp at the south end of Nimpkish Lake up to the north end, the site of an old loading works, camp L. Formerly, logs were transferred on to the log cars after their journey down the Lake for transportation by rail to Englewood. This was my route for getting from my office in Nimpkish camp to Dan's operation on Mount Holdsworth. After getting off at camp L I walked for miles along the abandoned railway right-of-way to my destination at Dan's logging – a round-trip of twelve hours.

On the way I walked between stands of young hemlock timber regenerated from the seed of trees still standing. Nature had rolled out a healing carpet of vigorous new growth, with no help from anyone, to replace the timber logged, made into rafts, and then towed to Vancouver.

So, on one day in 1962 I found Dan had followed his plan and converted the old rail grade into truck roads. He trucked logs from his active logging blocks down to his dump at Englewood. The former railroad operations had logged out all the timber accessible to rail transportation. Dan's logging operation was designed to reach out into the higher ground using log truck access. My job was to inspect his logging area, check that he had stayed within the marked boundaries, and had not skidded logs across streams or left good logs abandoned in the logging block. He must have had some twenty men working for him including a woman cook, all from Alert Bay or adjacent communities.

Dan needed three things from me. First, I provided him with an approved five-year logging plan. Second, I marked out on the ground the road extensions needed to reach next year's timber. Third, I liaised with the BC Forest Service district ranger at Alert Bay. The two recollections I have of Dan's operation in 1962 or earlier is the rock crew drilling and blasting rock for his new access road to reach the next year's cut. Then, his brand new 90-foot Madill mobile steel spar, painted yellow. It was a beauty and it revolutionized the way logging was done from here on in. My dates are approximate when it comes to discussing Dan. His log production was mainly hemlock-balsam.

My last memory of Dan's logging operation was a message I received while in my office at Nimpkish. A boomstick (66 feet long) had come down the hill in the turn and pierced the radiator, and bent some key brackets on the steel spar. That shut him down for two weeks.

Incidentally, families from Beaver Cove still occupied the married quarters at Englewood. Cec and I had to go there to service domestic fire extinguishers.

Marilyn Richmond (nee Livingston)
My future wife showed up in Nimpkish to teach school that fall. Roy Jewesson and family returned from Prince George to pack up and have a farewell party. Those of us in the forestry bunkhouse had looked after their dog, Twinkle. Since the teacherage needed some work first, Marilyn stayed with Nancy Jewesson.

Undoubtedly, by this time I had taken over Roy's job on an 'as need' basis. I also moved into a staff house in at Nimpkish at some stage. I remember seeing Marilyn for the first time. I was on my way to a slashburn when I passed Nancy Jewesson, Marilyn and Twinkle on the street. The school board had equipped the teacherage with a wood stove. Once in a while I got to split wood for her, and in return learn to eat pork loin for the first time. It was delicious. Eventually the school board switched out her wood stove for an oil heater. Funnily enough, I still showed up for Sunday suppers.

1962

Typhoon Freda came through the Nimpkish Valley on October 12. This altered the entire logging plan for the next year. Much of our proposed cut submission to the Forest Service was drastically altered to adjust to the acres of salvage needed. In fact, the division was in salvage for most of the next year.

The Nimpkish survey office was one-half of the forestry/survey complex. When I was first there, Leo DeHaan was in charge of the survey office; Ray Rucker followed him. Then Ray left for greener pastures, and my good buddy Harry Farmer took over in the fall. I have just chatted with Diane Farmer and she reminded me of the all-nighters preparing cutting permit maps for Head Office. From there, they ended up at the Forest Service district office in Vancouver.

Marilyn went back to UBC that fall. But Bill McMahan from head office had told me I could not go on my honeymoon until I had my Registered Professional Forester licence. So I spent considerable time that winter working on my professional report (Nimpkish thinning plots).

I have to relate one special occasion: I got to be the guide on a visit from Bill McMahan in his role of general manager. Anyone from Head Office was treated with awe. I remember him sitting in my pickup as we drove up H line[16] on the way to somewhere. He glanced up the hill to the top edge of the logging just on the Woss side of that patch of timber reserved for specialty products such as bridge timbers. He was fascinated at the green carpet of new forest high up on the sidehill – it had all been cold-decked years before. He noted that the Nimpkish would never have problems restocking after logging. Then he turned to talking of butterflies: how many wing beats per second did I think butterflies put out?? Lord only knows what prompted that off-the-wall discussion. I could not give him an answer. Weird. I went to Bill's funeral after he died, perhaps to get that answer.

1963

I had sent in our forestry budget to Ken Thomas at head office. It contained some throw away stuff but I got what I had been after – if only I remember what this was. Also, I had pointed out that I would be away on leave for 6 months (my honeymoon). Just before I left to get married, the budget came back with 6 crossed out and 3 written in pencil. A few days before I left to get married on May 18, I received my RPF designation. (Registered Professional Forester, No.403.)

Photo by Erik V. Peterson, Nimkish Valley, B.C. Tony And Marilyn Richmond, 1963

Shortly after I was back at work Kennedy was assassinated. I sat having lunch at home as that news came over the radio.

Sometime in that period Henry Hansen was fired and Howard Elder took over as division manager. That reminds me of an experience that is now folklore in the valley. One day after lunch I was driving down from Woss to Nimkish on the Nimkish road. It came through the Kaipit, past the iron mine, past the cut-off to Camp A. From there, it swooped downhill on to the long bridge across the Nimkish River, one lane fits all. I got the nose of my pickup on to the bridge itself, only to find that another pickup had come on to the bridge heading from the other end. One lane plus guard rails. I steered up on to the right hand guard rail and Howard and I passed without incident. He had thrown his vehicle up on to the guard rail on his side.

I do not think that performance was ever repeated. A legend was born. That time, speed was of the essence.

1964

This year started out as usual. Roy Jewesson was long gone. I believe Zoltan Fulopp had also left to fulfill other aspirations. Ray Rucker left and Harry Farmer took his place in the survey office. That Nimpkish survey office became a 24-hour day operation in the fall. This office produced all cutting permit maps needing Forest Service approval, five copies. These covered 4 million cubic metres of logs. They were coloured by hand, map legends applied, and were then sent to Vancouver for further processing. It was a batch process, but no one had time to stand down afterwards. Work as usual.

During one huge storm in November my daughter Nancy decided to arrive into this world. We thought the wet spot on the sofa was a contribution by the cat. However, recently I sent an email to my daughter Nancy to be read at her 50th birthday party in Regina:

"On Nov 6, a Friday, a foetus, full term, more or less, resided in its nest inside Marilyn. Sex unknown. Dr. MacLaren had told her that she should go over to Alert Bay to the waiting mothers' hostel the following Monday.

That evening at about 7 pm, Marilyn got up from the chesterfield at home. Tony noticed a wet spot on the cushion. He blamed the cat. But then reality struck.

Marilyn phoned Dr. MacLaren in Alert Bay. He said Tony could bring her over.

The weather that night was an abomination. The Nimpkish Valley on Northern Vancouver Island. just across Johnstone Strait from Alert Bay, was reeling from the aftermath of a hurricane.

But the neighbours rallied and a convoy started out for Beaver Cove over some logging roads. The trip should take an hour or so. Tony spent the whole trip worrying in somebody's back seat. Water gushed out of the sidehill and shot across the road in several places. Would the road wash out before they could get to the Cove? freak –

freak. Marilyn adds: 'Tony kept asking me how my mum was when she had him (Tony). He was out of it!!!.'

At Beaver Cove they waited and waited for the water taxi from Alert Bay. But it was too rough in the Strait.

Eventually, the taxi did arrive, and deposited Marilyn and Tony on the dock at Alert Bay. They finally got to the hospital at 10 pm.

Nancy was born the next morning around 5 am. Tony spent the night on the pool table in the hotel since it was totally booked up. This might be viewed as one alternative to a stable.

At 9 am, the hotel guy told Tony that he should go up to the hospital. When he saw Nana for the first time, he thought that he picked the wrong new born. They all had jet black hair. Alert Bay is also the home community for the Namgis First Nation. They are part of the Kwakwaka'wakwa tribe inhabiting BC's Central Coast.

Later that day, November 7th, Tony hitched a ride back to Beaver Cove on the forest ranger's boat. It was still blowing a gale. 'My job was to prevent the anchor from falling over the side as we fought our way across the strait.'

A week later, Nancy got her first ride in the company Beaver aircraft on floats. They landed at the airplane dock on the Lake at Nimpkish. Tony picked Marilyn and Nancy up and headed home. No sooner had she entered the house, Marilyn plunked Nancy down on the kitchen table and headed back out the door to meet her neighbour. They both went prancing off down the street. Tony has that picture embedded in his brain. Immediately, Nancy began to cry. Tony yelled out of the window, but they just laughed and kept on walking.

So there you are, Nancy Marie Richmond. Johnson. Gardner."

Ω

On December 10, I finally had my last cigarette after ten years of puffing away. I say this as it marks the period in time when Lorne Johnson, from head office, offered me the job of Vernon camp superintendent. Joe Bako took over my position as forestry manager in the Nimpkish Valley.

On December 13 at 10 am, on the old KO spur now converted to an extension of the Duncan Road, I came around one of those sharp curves bordered by a steep bank. I was on my way up to meet Ole Jernberg, the guy I was to take over from as Vernon camp superintendent. I met another pickup hurtling around the corner from the other direction, so I steered right up into the base of a vertical bank and missed Art Newman by a hair. I thought that I had better back up to see if Art was OK. There he was in his pickup on the road, white in the face and knuckles, and staring straight ahead. I reached into his pocket to roll a smoke. I think I took a puff or two before I remembered that I had quit. Art was in charge of all falling and bucking in the division – a big job. That was my last cigarette ever. Once I started at Vernon, I told everyone I was not a smoker, and this made my quitting easier. Art was probably production superintendent by this time and he became my boss.

Some will remember that winter because of the intense cold and four feet plus of snow between Woss and Vernon. This brings to mind the road through the Nimpkish Valley was now extended down to Gold River. Marilyn and I and our brand new baby daughter, Nancy, joined a convoy of vehicles that left for the south at Christmas. I drove Eric Cooke's Volkswagen beetle since he did not yet have a driver's licence. Eventually we got bogged down, and then were towed on top of the snow to Campbell River behind a pickup with 4-wheel drive. The Volkswagen beetle tobogganed on top of the deep snow the whole way. I drove the beetle into Vancouver and left it parked at its destination.

Ω

Chapter 9:

Nimpkish Valley, Vernon camp, 1965-67

Figure 13 Steel tower set up to yard logs. Most of ours were on tank retrievers, rubber tires. By that summer, Pat and I looked after five of these. Photo source not identified.

The scenario

I have great memories of Ole. His background was as a highrigger, preparing spar trees for logging. Going into Vernon camp was always a pleasure and Ole helped get my duties completed every time. The last time I saw Ole was when he was walking home across the bridge over the Nimpkish River between camp and the married quarters. A large man followed by two little Schnauzers.

93

I assume that Vernon's 1965 target production was 1.127 million m3 or 225 million board feet. We went from three towers in Ole's day to five plus two snorkel Washington track loaders to top it off. Ole had parked all the equipment for the Christmas shutdown at Sutton reload. It was frozen up. After Christmas, as soon as the crew returned to work, we were to move 0903 down to Woss to start logging on one of Jack Vettleson[17]'s spring shows. Over shutdown some four feet of snow accumulated on our own logging areas. We lit a fire under the engine to melt the steering oil so it would steer. Don't forget – I was the brand new logging superintendent, and stayed out of the way. Boy. Was it cold! The crew did wonders in severe conditions.

On the way down to Woss, the steel spar 0903 on its tank retriever platform rolled into a 25-foot deep ditch after sliding off the road due to sheet ice. I think it rolled twice. No one hurt, and we even recovered the operator's lunch kit. Once rigged up, we got a maximum of 6 logs a day due to the deep snow. The chokermen used shovels to dig out each log so they could attach the chokers. We should have hibernated until spring. However, it was boom times in the industry so any volume at all was a bonus. How Paddy Smith got the trackloader (rubber-tired on wheels, not tracks) down to Woss without skidding off the road I'll never know. Actually I do though. That man had nerves of steel and took that machine through the curves at top speed. That way, he stayed on the road.

I have to refer to Paddy once more. He was a Royal Navy diver before coming to Canada. At Englewood, he ran the steam crane used by the railway to build and repair bridges, as well as attend to train wrecks. When Ole received a Washington trackloader complete with snorkel, Paddy got to operate it. He was awfully hard on equipment though. However, he was a key component in achieving our daily log production.

The staff

Patrick Conway was my woods supervisor. He had expected to have the superintendent's job and was deeply hurt that a wet-behind-the-ears intruder had encroached on his turf. We had it out on a brow log way up in Lukwa Creek. I had an image of me picked up by the seat of my pants and tossed down into the slash. But we talked and made a deal: in return for him teaching me how to log, I was to teach him management skills. Shortly after I got into my new position Howard Elder was fired and Owen Hennigar took over as division manager with Art Newman as production superintendent. So, I was surrounded by loggers. My problem? I had two: Art was a micromanager and could not stand Pat, and Pat had no use for Art. This came to haunt me throughout my time at Vernon. And I believe it did much to sway the judge in favour of Tahsis at the conclusion of the Muchalat fire court case – more of this later.

Jack Van Graven, the bullbucker, was a true professional. He and family had come from Holland at one time. We lived beside them. Bud Lancey, ex American and his wife Maxine lived in the married quarters down the road. Bud was the construction boss – roads mostly. I was so green that I left all road work up to him; except for the crappy gravel he put down on some of the new road. Of course, we were limited by the characteristics of our available supply; but, why should all the ball bearings appear on the steepest pitches and on the corners! Mrs. Lancey operated the store/coffee shop. The Lanceys brought the first private car on to the claim once the road was opened up to Gold River. A large red Dodge sedan.

Our master mechanic supervised the mechanic's shop. But he was not my master mechanic really. He was supervised by Holger Wickstrom, the mechanical superintendent in Nimpkish, 80 miles away. So our mechanical services were not ours to manage. The mechanics did most of their work at night after we had shut down for the day. On one occasion I was at the Sutton shop. It was in a disgusting mess. I picked up the broom and started to sweep. The master mechanic implored me to quit doing this. But I never did get to see this shop looking clean and tidy. This told me a good deal and did much to maintain my uneasiness with the mechanical set up in camp. We had millions of dollars in parts and stuff at the Maquilla warehouse. I refused to own a key to the place – the fewer people with access the better. We did have a good warehouseman, but eventually he was kicked out of camp because of erratic behaviour due to drugs and alcohol.

Back to mechanics. On one night shift the mechanics changed oil and filters in the 0905 tower up on the side of Vernon Lake (East

95

side). The new filters were installed upside down and that cost us an engine and production. Also night shift welders were a real danger during fire season. We fought more than one welding fire during my Nimpkish years. We had a welding fire up on that Vernon Lake sidehill while I was superintendent. I always sensed a lack of support for our mechanical needs while I had that job.

Management and administration

On one occasion, the steam plant went out at camp. It supplied heat and hot water to the bunkhouses, cookhouse and the office. We had recurring troubles with that plant such that clothes were still wet the next morning. This was January – cold and wet. Add in cold bunkhouses and other annoyances connected to loss of heat and hot water for showers on a wet, cold January day. This interrupted cookhouse operations as well. The steam plant was the responsibility of the mechanical superintendent at Nimpkish – and since we were 80 miles from him and his crew, out of sight out of mind. Anyway, after the boys went out on strike one day due to the lack of dry clothes (I joined them), we got some attention, although at the loss of one day's production. The mechanical superintendent's negligence was noted. We suffered significantly from a lack of support services in general. We were so far away and communicated via a 9-gauge phone wire to Nimpkish. The phone line wound from married quarters to camp offices before it got to us. Signal strength was weak and anybody could pick up the phone and hear the conversation. Very primitive.

Our timekeeper or administrator distressed me on many occasions by sleeping on the counter. He answered to a human resources supervisor based in Nimpkish. I knew he was bushed and managed to get him transferred back to Nimpkish.

We had a radio system that connected most of us throughout our operating area – except beyond Sutton reload. Radio reception was dead from the area between the end of the railway mainline at Sutton, and down into the Oktwanch River system that ended in Muchalat Lake. We lost a whole day's production on the 0907 (Tom Holo) when an O ring blew. That crew sat there broken down until I discovered them that afternoon. Also our cut plans for this area were badly jumbled up between the Tahsis limits and ours. We were logging down into their territory and vice versa.

96

The cookhouse was another part of the job. Marilyn finally put up a sign at our front porch to say, 'This cookhouse closes at 6'. I ate more in the camp cookhouse than at home. One year I went through 7 cooks and 13 bakers. When we pulled out the flour bin after getting rid of the last baker, we found it full of empty whiskey bottles.

We never had enough skilled men. I recollect us pulling a hooktender out of the ditch south of Sutton reload. At least he said he was a hooktender. It was a Monday morning, and he was on his way back to a Rayonier camp close to Port McNeill. He ended up on one of our towers, got slapped across the head with the haulback. He was flown out to Campbell River hospital. Safety was a big thing for us though. Owen Hennigar worked on us to work on the crews to work safely. My other shortage was rigging pullers (rigging slingers).

Terry Dick

I do not remember his real name. His home was also in Alert Bay. We were desperately short of skilled loggers. Our camp was the furthest from the coast, and the largest with 250 men in the crew. Turnover among the skilled loggers was high. We became the training ground. The hunger for logs was great; we tried to keep the staffing levels up so wood flows were maintained. The sawmills and pulpmills down the coast were truly hungry. Vernon camp itself had recently switched to trucks from rail. Our greatest lack was a reliable mechanical repair facility and trained loggers.

Highlead logging with steel towers required a crew of nine men. One key job was that of 'rigging slinger'. His main task was to work with his two chokermen hooking up the logs in the cut block to the mainline for swinging to the landing. We had five towers at Vernon camp and needed 45 men to staff all operations connected to putting logs at roadside for transport to the reload. Terry showed up one Sunday night after a journey from Alert Bay by speeder on the Nimpkish mainline to go to work next morning as a rigging slinger.

The foreman reported to me later that Terry was just another 'wannabe' rigging slinger and that he was learning on the job. My gut feeling was that Terry had worked for Dan Hanuse previously as a chokerman. Logging crews generally had to quit one operation to hire out to another in a higher job category. Anyway, Terry's absences and unreliability ended his relationship with Vernon camp. In 1967 I also left the Nimpkish Valley to return to university. After ten years in the backwoods my mind was frozen shut.

The lesson from our relationship with Terry Dick, however brief, was that Indians did not work well at a distance from home. I found later that seven days away from family was about the upper limit. This theme reoccurs later in my journey with Indians.

The Nimpkish Valley was a route in earlier days for Kwakiutl people from Alert Bay on the northeast coast of Vancouver Island to visit Vancouver Island's west coast. In turn, Nuchatlaht Indians in Zeballos on the west coast of Vancouver Island visited the Nimpkish Valley for many purposes. I was told that Klakakama Lake (south end) in the centre of the Nimpkish lands was used by First Nations en route to their summer camps. This was a final resting place for their old people prior to death. It was a peaceful location, almost a small meadow, surrounded by cottonwood trees. On the way back home, the families conducted the necessary ceremonies and disposed of the bodies according to their customs.

Loggers had discovered many cedar trees beside the middle reaches of the Nimpkish River over the years that were split open, then abandoned because of defects, by Indians building canoes.

Ω

My only other accident was down at Maquilla reload. We were about to close operations for summer shutdown. Just as the last load of logs was lifted up off the log truck then down on to the railcar, the peaker log dislodged and fell on to the head loader as he bent down to release the slings from the gun barrels. It was serious. I checked with him later in hospital, but his back was broken. He was slated for a wheelchair the rest of his life. I often think of him and his family. Normally, much care and attention is paid to those peakers. They are dangerous, and can roll off and hurt someone; although the load bridles prevent this very thing in theory. But, someone at some stage has to remove them in order to unload the truck. The last truck unloaded in Vernon camp before summer shutdown.

Pat Conway had an Irish temper. He was always on the lookout for those trying to pad the timesheets. That weekend, the lowbed had moved much equipment, all on overtime. On Monday morning in the marshalling yard, Pat took the lowbed operator's timesheet and tore it up in front of the assembled crew (our payroll was 250 men or so). I made him apologise in public and sign the timesheet, now somewhat crumpled.

Safety for fallers was an ongoing matter. In the summer, they worked in smaller timber on steep sidehills, but in the winter, the valley bottom timber fooled them. Big branches would fall out of the tree crowns (widow makers); in some cases a faller would get greedy and forget safe work procedures. A few were hurt or killed. I became really concerned each fall as they moved back into the valley bottoms and big timber. Jack Van Graven though was diligent in his monitoring of each man. We had 46 fallers in camp. When I became super I was told that these guys would be my worst nightmare. So many of them were commies and left over from the IWW (International Woodworkers of the World, or 'Wobblies') days. However, I lived by the current IWA (International Woodworkers of America) contract, and within a year the hard cases had left and taken up residence at MacMillan Bloedel, Franklin River camp. I used to hear of rumbles from that direction once in a while.

Fallers with back problems were a constant concern. The Workers Compensation Board (WCB) fought mightily against fallers' claims for back problems. It was so hard to present medical evidence; yet, for someone with a back injury the discomfort was remorseless.

Ω

I must record the 0905 incident. The boys were logging a setting at the edge of the Nimpkish road down towards Muchalat Lake. A tailhold pulled on the yarder and the tower fell over on to the loader. It landed an inch away from the operator's cab. Phew. We had to remove a bent section of the tower and replace it. How much production did we lose from that incident? We could have killed Fred Laliberte, the loader operator. He was Métis. I found out years later, during my Saskatchewan sojourn, he was probably from Ile a la Crosse, Buffalo Narrows or Beauval – all of them together representing the heart and soul of the Métis kingdom in northwest Saskatchewan. He was a reliable worker.

One summer, we got Patrick and Loretta to go away for a holiday. We were both working steady and never got time off. But Pat did leave me for two weeks to be foreman and everything else. During that time we moved and set up the towers twice without incidence. Pat had kept his end up by teaching me to log.

Just two more real irritations, the latter more serious than the first.

The accounting system.

Vic Van Slyke, the accounting manager for the division, had set up a new accounting system that coded everything right down to the teabags used in the kitchens. In principle, it was an excellent management tool. I should have been able to have this at my fingertips. For instance, how many board feet produced on a $1,800 mainline? I never knew; how many chokers per machine per one thousand board feet (MBM) of production? I never knew. We did not receive these statements until 6 weeks after the month in question. We superintendents sat around the table in Owen's panabode office at Woss discussing all the numbers. It was just all too late for me. During those 6 weeks I was into a whole new set of problems/issues. I felt so inadequate around that table. And add this to Art's poring over every nickel and dime from situations 6 weeks old; it only deepened my sense of inadequacy. Woss had people in the office that could do the analysis, but we only had one guy sleeping on the counter. Pat and I were totally overworked and understaffed to develop rational thinking levels. With 5 towers and two snorkel-equipped Washington trackloaders, we had 7 effective landings in play at one time. Owen remarked to me after one of those meetings that he wished he had my education. I, in turn, told him how much I would like to acquire his logging skills. Next to Roy Jewesson, Owen was a person I respected deeply.

The second foreman.

– I will call him X since I have a mental block about his name. Anyway, X came to us from Spring Creek. We split the towers up and X took some (the ones closest to Woss) and Pat took the rest. Now X lived in Woss and commuted daily to Vernon. I failed to integrate him into our management team because he was never in the office to sit down with my other staff to plan, forecast and shoot the shit. I had ended the office beer drinking on Friday nights after I took over. You cannot go after drunkenness in the bunkhouses if you are doing your own tippling. Given that we had to plan equipment moves, where to put the next landing, how many boomstick cars and where should they go (especially when you have 2 reloads working), has Bud built the necessary log landings.? This needed real integration at the management level. But X returned to Woss every night, where Art Newman and he tried to run our operation. I got these evening phone calls second guessing all our decisions. Then I would get balled out over the 9-gauge phone line with everyone listening in.

I almost caught it one day at a tower set up on a landing just above Maquilla reload. The fairlead at the top of the tower broke off and thudded to the ground not two feet from my right boot. Within days all sides were towered down and the fairleads and bushings inspected. We followed this with greasing every pulley and bushing. Normally, this is done as the machine is moved and the tower raised back up. I was also insistent that all loading machines were inspected for cracks, seized pulleys and wire rope conditions regularly. The Washington Trackloaders and their attached snorkels received a lot of torque. We finally reinforced them with T1 steel plating. Our Washingtons were rubber-tired and not tracked.

Paddy Smith operated one Washington and Albert Mercier the other. Pat spent hours nurturing and mentoring our younger loader operators. One youngster had a sore peepee until we pointed out that holding it with a hand covered in diesel was the problem. The other young fellow picked up a mainline with his grapple loader. We were set up on the flats close to Maquilla that time. We wrote off one almost new mainline.

The tours
We had two visitations that I recollect some of:

First, a full dress visit from head office. I picked up the convoy at Croman reload and brought them up to Vernon via the Nimpkish Road. As we came by Klakakama Lake I told a story over the radio.

101

We were following the annual migration route of the Kwakwiutl Indians from Alert Bay over to their West Coast hunting and fishing locations. I made up a translation of the word 'Klakakama' as meaning 'sagging bosoms' in the Kwakwiutl language. This little gem of information was greeted with a dead silence.

Second, John Liersch, ex Powell River Co. joined head office. Later, he and Owen showed up in camp in one of the two red Pontiac cars kept at Nimpkish. John wore that familiar blue and white striped work shirt favoured by machine operators and other woods personnel. It was a sign of being 'in the game'. The stripes were narrow and those shirts wore forever. After his visit I started wearing such a shirt. Anyway, the only part of that trip I remember was that he was disgusted at the rubbish on the ground around the coffee shop and phone booth. I was told to get it cleaned up pronto. We had stopped at the 0907 set up down on the Oktwanch road. I did not let on that I had found a full bottle of whiskey in a culvert just down from the tower.

Muchalat fire, 1965.

In the fall of 1965 we were slashburning settings south of Sutton and on the Alston Road. We had the Nimpkish road extension hooking up with Gold River (Oktwanch Road) and were logging intermittently on settings down into the Oktwanch. The timber licenses in play were in a quilt pattern across the Oktwanch valley. Canfor and Tahsis operated alternate licences; although, technically, Tahsis was operating inside TFL 37. We drove through a corner of Tahsis TFL 19 to get to some of our timber, and vice versa. Valiant efforts were made by both companies to conduct timberland swaps. This was not to be. Bill Ford of Tahsis had logged on the west of the Oktwanch River, while we operated in our limits on the East side.

I had set up a weather station at Holiday Creek – sticks, max/min thermometer, wind, and Relative Humidity readings. I took the readings several times per day. Our slashburns that fall were not that easy to light and we achieved moderate slash reduction. Where the terrain steepened towards Muchalat the results were more complete. We finished burning in the first week of September or so. The higher up the sidehill, the drier the slash. We had one setting where we got a reburn. We used two fifth-wheel log trucks that carried 4,000 gallon portatanks. These stood at Sutton reload during

102

fire season, on their stilt legs. We ran hose on the top side of this setting and had crews mopping up here and there. By September 19, we had reduced our mop-up down to one crew with a fire truck on patrol. On September 20, I asked Pat to bring some of his logging crews after work to extinguish some hotspots along the Oktwanch Road, together with the portatanks transported on fifth-wheel logging trucks.

Owen Hennigar came along that evening and told me to send the crews home so they get back to logging the next morning.

At 5.30 am on the 21st of September, I conducted another inspection of the Oktwanch Road. Unknown to me at the time, Pat had visited the Tahsis Co. loggers on their side of the Oktwanch River at 0430 hours that morning. The fire was about to jump across the River on to our side. He went over on to the Tahsis side of the River to pull hose and help them fight fire. Tahsis had only started their slash fires a day earlier since their slash was drying out slower than ours due to an east facing aspect. They were in shadow until later in the mornings.

On the morning of September 21 a cold front passage with jet stream winds hit the Oktwanch Valley. They funneled down that Valley and blew the fires into a campaign fire from one end to the other. Do you remember Roy Jewesson's discovery that relative humidity in front of a moving fire is zero percent? Well, that is what was happening by 11 am that morning. I heard later that spot fires came within 8 miles of the Tahsis pulpmill. Pat had returned to Vernon camp by marshalling time at 0715 hours. But with that wind he recommended that we send a fire truck and a fire patrol down there immediately. I followed the patrol down there later to see if we should dispatch more crews in support of the fire patrol. Do not forget that we had no radio contact there.

Later, I followed the mopup crew with the fire truck down to the junction of the Oktwanch Road and the Tahsis road. This connected with the Tahsis road system on the west side of the Valley. By 1 pm I had reached the north end of Muchalat Lake. Tahsis was logging with a skyline rigged tower, a very special piece of equipment, on a ridge just above the junction on their side of the

River. Tahsis removed it out just in time. I drove up on to that Tahsis landing, where I could look right up the Oktwanch Valley. I faced a thick wall of smoke flowing down and out over Muchalat Lake. But this wall of smoke did not travel vertically, but rather horizontally. The horizontal sheer wind prevented it from rising. I have never observed that phenomenon since. My position on that landing was above the smoke. I looked down. That roiling, solid wall of smoke flowing down the valley resembled molten magma flowing out of an erupting volcano.

I then returned down on to the flat at the head of the Lake. There, I was below the smoke stream. I had the fire truck standing by in a gravel pit. We noted the fire was at canopy level. In this 800-year old mix of Sitka spruce, Douglas fir, cedar and hemlock, there was no wind at ground level. However, burning clumps of lichen were descending gently to the ground and igniting the forest floor fuels, then whooshing back up into the crowns. It was a staggering phenomenon. We left the fire truck – a yellow cab with the tank painted red, but with no windows, parked in the gravel pit at the road junction and got the hell out of there. As we travelled back up to Sutton reload, flames were shooting out from the timber edge on either side, just like a flame thrower pointed directly at us. I must have rescued the weather station en route. As I returned with the fire crew, the whole valley behind us blew up. That night we had all the Tahsis manpower and equipment parked in Vernon camp and out at Sutton reload.

The next day, Sept 22, we had air tankers dropping bentonite[18] up the Alston Road area. The Martin Mars side-dump air tanker came in a week later, once the smoke thinned out a bit. Tahsis looked after the south end of this massive fire area and Canfor (me) took over the north end. The fire left nothing in its wake, just rock. No duff or humus to speak of. The standing timber on the high side was scorched down to the heartwood. Roy Jewesson was quite right: fire creates its own climate. One morning, Art Newman and I clambered up one sidehill into the timber just as the Martin Mars side-dumper let loose a load right over us. It was like walking into a stand-up refrigerator. The tanker was filled with sea water mixed with a gell or gum made from seaweed and cellulose (my recollection). If that had been the belly dumper, it would have been 'curtains' for both of us.

One month later I returned to that gravel pit and there it was – the fire truck totally pristine in its red and yellow colours. Everything else from Sutton reload to Muchalat Lake was burned to a crisp. Fried trees sitting on their root systems with everything else burned away. Wind up their skirts, so to speak.

In those days we knew nothing about cold front passages, jet winds and fire behaviour indices. We had nothing to forecast them with anyway. We had our sticks and relative humidity meters. In Vernon camp I do not remember ever seeing a weather map or forecast. Pat was still in his woollen Stanfield long johns. We fought that fire for a month.

Ω

Vernon camp, 1966.
We got a period with icy roads early in the year. I was in my truck by 0400 hours with bags of salt. I sprinkled the salt on active truck haul routes where a steep pitch, all iced up, could delay log hauling immediately. I loved those 4 am mornings, no radio chatter. Once in a while I greeted Jack Vettleson, Woss camp super, on the radio, doing the same thing.

On another occasion, at winter shutdown, someone taught me to operate a D8 cat; I plowed snow to keep camp access roads clear. In reality, it was an amateur performance. And then, during summer shutdown, just a small construction crew and me. Once, the woods foreman from Woss came up and taught me how to splice strawline.

More recollections
One early morning when I drove into camp I got the shock of my life. All our transportation, 18 vehicles parked in the marshalling yard, were swept on their sides and all piled up at one end. How did this happen? A night train with empty rail cars passed us as it headed for Sutton reload, just up the track. On one car, someone had loaded a bunkhouse on skids – sideways. It jutted out enough to catch the row of crew buses, including the ambulance, parked overnight. What a

disaster; however, the crews went to work, some a bit late. I went over to the payphone, contacted head office in Vancouver, and alerted Robert Bentley, the guy in charge of the Company's insurance. For that, I was chewed out by the mechanical superintendent at Nimpkish for going over his head.

Rarely, a February can reach fire season levels of drying. Sometime in that February, I sent out a fire patrol, and lo and behold, they found a fire at roadside, and extinguished it. I received a pat on the back from head office (Tom Wright). I had learned this lesson from my fire warden days. We put out a bridge fire on bridge 16 – in February. If that fire had weakened that bridge, the whole logging division would have shut down. Also, more than one cold-deck burned down in February on the Coast.

Ω

I adopted snorkel logging with trackloaders within cutblocks by bulldozing access roads in herringbone patterns. One day, we set a division production record. The boys in the bunkhouses were celebrating. For once, I joined in. Unfortunately, my wife's lifelong friend and her husband drove into camp that afternoon. We did not get visitors as a rule. I showed up late at home, three sheets to the wind, then collapsed into bed. And they had cooked supper. I do not think I impressed anyone present.

My daughter Kathy was born in October, on the long weekend. Marilyn and I were leaning over the bridge on the Oktwanch River watching the salmon spawn. This appeared to precipitate labour. A quick trip to Campbell River came next.

Someone shot at a faller as he was working on a cutblock south of Klakakama Lake.

The Woss master mechanic almost lost his life as he drove up the Nimpkish road at night and met a washed out bridge just on the Woss side of Maquilla. He was unhurt, just shook up.

The break-in at Maxine's coffee shop. I became a witness at the trial in Campbell River. Roderick Haig-Brown was the magistrate. I failed the test. The guy was declared not guilty.

My daughter Nancy, at 18 months old, came face to face with a bear at her sandbox beside the house. Marilyn took the broom to it and away it ran.

As superintendent, I had little recreation time. However, on one Sunday, Marilyn, Nancy and I took out our little outboard to the south end of Vernon Lake for a picnic. Later that afternoon the wind started to blow from the north. We had a whole lake in front of us as we prepared to travel home. The waves were high enough to cause me notable concern. Each time we went up and over a wave, the transom met the waterline. Anyway, that was the last of our boat trips. That trip scared me deeply.

Vernon camp, 1967

In August 1967 up on Vernon Ridge, I gave my notice to Art Newman, mainly because I could see that he was unhappy with me. He had the seniority and had been in the valley a lot longer than me, so I got out of his way. I understand that the powers-that-be were truly exercised at having to pay out so much since I had so much unused holiday time. I returned to UBC and started a Master's degree in forest management.

$$\Omega$$

Postscript 1968.

I was writing my final exams at UBC in April 1968 when Tahsis and Canfor met in the British Columbia Supreme Court to fight over damages left behind after the Muchalat fire. Prior to that, Canfor called me into a Vancouver boardroom to meet with Canfor staff from Englewood. We conducted an in-house mini examination for discovery lead by lawyer Dumoulin. He was otherwise busy with a large court case in Ottawa. He showed up at our meeting smelling of sherry. I left this meeting confused. Art Newman refused to consider putting Pat Conway into the equation, the very witness to Tahsis' slash fires exploding and crossing on to our side of the Oktwanch River.

Soon after, I stood in the witness box as a witness for Canfor at the trial. I spent two days on the witness stand and was effectively roasted by the Tahsis lawyer. I should have had my own lawyer. We would have probably won the lawsuit. The Tahsis lawyers were loaned my weather station data and conveniently lost this information before I could review it.

How did Tahsis manage to drive all of its high-priced equipment up through my fire area and park it at Sutton reload? Their own timber was burning so fiercely, their road system was impassable. It was not possible to take it back to their camp through their fire area because of extreme fire behaviour and temperatures approaching 1,100 deg. C. Canfor paid me $50 per day for two days. I returned to my exams and just squeaked through a graduate course in cost accounting.

I tried to stay out of Faculty of Forestry courses as much as possible. My supervisor, Professor Harry Smith, did let me take a graduate course through the Faculty of Commerce – Organisational Behaviour and Administration. I received a final mark of 97%. But the power I gained from being able to articulate all the management/administration turmoil in the Nimpkish was truly rewarding.

Amidst all this turmoil, our son Graham was born on February 9. Another one of God's gifts. One that did not take hours and hours travelling over logging roads to collect.

In conclusion
My Nimpkish years are irreplaceable. I met Marilyn, I learned hugely and the experience has served me well. I have to thank my professional godfathers, the Prentices and Bentleys for my invaluable education on the job. It was a great team at Vernon camp. Patrick, I have always missed you since.

Ω

Chapter 10:

BCIT

I found a job teaching students in forest fire management and forest measurements at British Columbia Institute of Technology (BCIT). I found two excellent mentors: Eric Crossin, who taught me to prepare lesson plans and Alf Cuthbert, who showed me how to set a fair exam. Prior to this, BCIT dispatched me to a forest fire simulator course at Northern Alberta Institute of Technology in Hinton, Alberta. Totally fascinating. On our final evening, we were served buffalo burgers – my first. Back at BCIT the initial focus was on fire simulator training for industry and the BC Forest Service. The course leader at that training course in Hinton, coached by Alberta Forest Service fire people, advised us never to have management in the training session with us. He considered management's presence far too stressful for the participants in a fire simulator training course. Rubbish! How can you fight forest fire without the input of your boss?

Forest fire simulator (1)

Forest fire simulation became a key component of my 'wardrobe' over the next several years. I thrived on it. On one occasion, we fought fire by satellite. My team was in place at BCIT, with other fire teams set up at Kamloops and Prince George.

Back at BCIT our forest fire simulator was set up in a special classroom. There was a bank of six overhead projectors installed in a large cabinet, as well as a 35 mm projector. The projected image of the assigned landscape projected as a large diapositive. This image projected on to a screen that divided the student participants from the simulator operator. This screen supported rear projection.

The cabinet containing the projection equipment held the five projectors that beamed light up through the tabletop on to glass panels. From there, lenses and mirrors threw projected light beams on to the rear of the screen. The simulator operator was able to project the landscape to be used in the simulation exercise. By painting the glass plate black, the operator used a sharp instrument to remove the paint to create catguards, fire front, wind direction and flame front – in response to interaction with the fire team.

One overhead projector for each layer. The smoke projection had a smoke wheel installed placed to control intensity and direction. As did the fire wheel. It was all realistic.

Add:

-Project flame

-Project smoke

-Alter wind direction according to the script

-Fireguards and noise: helicopters, trucks, caterpillar tractors, the sounds of a travelling fire.

-Complete telephone switchboard with a radio communication system.

Sophisticated cooling kept all the overhead projectors inside the cabinet at normal temperatures. Trainees connected with each other through a telephone exchange and a radio. Helicopters and truck traffic had soundtracks and could connect with the trainees.

The simulator operator was 'God' who controlled fire behaviour and the success of fire suppression.

In front of the screen
Since forest fire simulation relied on role-playing, various participants filled the positions required by a real life industrial forest

fire incident. The two most frequent causes of ignition were lightning and spark fires that originated from malfunctioning logging equipment. The options also included the odd fire triggered by a dropped cigarette and a fire caused by welding. In British Columbia, most of the larger companies record their policies surrounding forest fire prevention and control through a fire protection plan. Within these plans, management has two potential scenarios: within working hours and after working hours and weekends. Remember, these were the days of logging camps, remote locations and primitive communications. I always included head office in the role-playing. We would put this individual in a separate room. Cell phones were not even thought of in 1969.

Briefing for the role players included weather conditions, time of day and day of the week. The role players noted availability of streams for water sourcing, also the availability of back up. Thrown in for good measure was a map of the potential fire area compatible with the landscape projected.

Forest companies sent in their logging camp staff, their fire protection plan and a landscape photo of an area they were familiar with. I then converted the photo to a diapositive image for use in the exercise(s). After everyone settled, I projected the smoke onto the screen and we were off and running. In all my years as God in the simulator – 20 in all – no one ever challenged fire behaviour displayed or the circumstances uncovered as the role-playing team involved themselves in controlling the fire underway. We followed each exercise with a critique in which their particular fire protection plan was thoroughly tested. Once each fire was projected on to the screen, the fire management team treated it as if it was the real thing. They responded viscerally.

I have a few favorites, one of which I shall relate. There were several over the years: A BC Forest Products operation from the Island sent over their logging overhead and key people to participate in a simulator training exercise. I set a fire up on a sidehill that involved a catside logging show. It was 1730 hours and everyone had gone home. It was also a Friday evening. That crew slipped into their firefighting jobs instantaneously. The catside foreman was a key part of the fire suppression strategy. I had lived through this predicament myself on more than one occasion. I knew the fire would escape uphill before the completion of a cat guard. On the screen, the cat guard wended its way uphill steadily in the attempt to cut the fire off from more forest fuel. Of course, the fire escaped uphill into the timber. He lost it during the exercise. He swore, had a tantrum and took himself out of the exercise. The camp super, though, immediately recognised the dilemma and, there on the spot, proceeded to organize a night shift suppression effort.

Simulator training was an ideal learning tool for second year forestry students at BCIT. We set up a project to build forest fire protection plans based on specific inputs and examples of industry plans provided. Then at the end, we took one of the better fire protection plans and used it as a fire simulator exercise. After this, the plans were returned to the students for corrections and fine tuning.

The Forest Service fire control staff based at Green Timbers came in to use the simulator. I set them up with the scenario, played God, then left. They would not let me participate in the debriefing. Insecure? Over the years, I have seen it all:

*Do you remember my involvement in the Woss bridge fire alluded to in Chapter 6? We burned that bridge down due to the lack of a gasket in the suction hose.

*Another of my favorite mistakes, observed more than once. Someone reported a fire. A supervisor went out in a helicopter to have a look. He came back to organize an initial attack crew. By the time the crew got to the fire it had blown up into campaign fire. The missing link? If that supervisor had some firefighting equipment onboard when he had first seen the fire, he and his helper could have extinguished it.

*Early in May 1984, a smoke showed up northeast of Prince Albert, SK. It was near Narrow Hills Provincial Park, north of Smeaton. Fire control staff from Prince Albert went out for a look. It was a long helicopter flight from the airport. They did not take a monsoon bucket.[19] That fire blew up and covered 60 or 70 sq. km before the rain put it out months later. They should have landed and commenced initial attack. One hundred gallons of water from the bucket would have saved the province some $750,000 in firefighting costs.

*In Ontario, I set up a simulator fire for my Lakehead University students in 1984. The favorite mistake there is to send out an initial crew complete with hoses and a pump by helicopter. As they are setting up the pump, the fire travels away from them. By the time they get water to the fire, it is out of the reach of the hose lay. Later, you will see that they should have cut a fireline and snipped off the frontal fire movement with pulaskis and chainsaws. A fire cannot spread if you remove the fuel.

Other fun times
My years teaching at BCIT were perhaps, the most rewarding component of my forestry career up to that time. After Eric Crossin and department head, Victor Heath were seconded to a forestry school north of Nairobi, those of us staying behind commenced experimenting with teaching technique. We pooled our classroom and lab time to convert to a sequential curriculum. I started out with a week of timber cruising and mapping: Bunsen Lake, Port Moody, in the fall, and Manning Park after Christmas. Dave Holmes and his forest engineering class followed with road layout, setting boundaries, bridge design. The logging instructor would then lead the class in logging and transportation, not to forget road construction and landings. We pooled the instructor team, each assisting in the various duties performed. The forest management instructor then wrapped up a project with detailing government paperwork required. Students were exposed to insect and disease parameters by the pestology instructor as they conducted their other duties. It was a successful teaching mechanism.

One more story. At the year end, all instructors involved in second year programs, sat together and presented the final marks for each course, for each student. Dave Holmes had one student that was so far superior to any he had taught before that he gave him 98% in forest engineering. We asked Dave to consider a 100% grade, and then talked Dave into posting a grade of 105% to reflect this young man's outstanding performance.

I had one student in the second year forest mensuration course, let us call him Malcolm, who received 44% as his final mark. Now, the usual protocol was to adjust the grade for a marginal student where the mark was close to the 50% passing standard. The second year instructors all sat in one room and presented the term results, course by course. In Malcolm's case, I could not raise it up to an adjudicated pass – the gap was too large. So, I failed him. He returned the next year and the same result occurred. I felt wretched, and Malcolm was angry. My policy was that 47.5% was the bottom limit for an adjudicated 50% passing grade.

Ω

In passing

In some ways, the BCIT years were truly rewarding. Preparing young folk who are hungry to learn was awesome. The sabbatical summers, wherein we were encouraged to stay connected with industry, took me to the Cariboo, Okanagan and Prince George regions. Field projects on the Coast and on Interior snow produced first class learning environments.

During one summer sabbatical to the northern Okanagan, my colleague Roger Cannon and I stayed at a summer cottage owned by a Forest Service fire guy. This reduced travel expense significantly. The cottage was on the lakeshore right across from Kelowna. One night we had a firsthand view of airtankers dropping retardant on a bush fire in steep terrain just across from us.

Perhaps though, my most poignant memory is of a mother duck plus her crew of very young ducklings swimming slowly past us in the reeds at the lake edge. Everyone was chirping and feeding. The next morning, with a wind throwing up small waves from the south, the ducklings retraced their route – but without mum. They would not have lasted long after that.

Ω

Later, UBC forestry students with their fire professor, Robert C. Henderson, also fought fire on the BCIT simulator. This connection leads to the creation of International Forest Fire Systems Inc. (IFFS). See Chapter 11 to follow.

Ω

Chapter 11:

International Forest Fire Systems Inc. (IFFS)

Figure 14 IFFS logo

Note:
The 'sky spider'[20] system delivered highly trained initial attack crew to a remote forest fire. The purpose was to put out a remote lightning fire before it spread. This is the first documentation of IFFS. Handpicked students from both UBC and BCIT spent the summer going down a rope from a hovering helicopter. The balance parachuted to fires in the NWT and the Yukon.

The BC Forest Service adopted this system in 1978 and changed the name to 'rapattack'. The Alberta Forest Service stole the system at about the same time. In Saskatchewan, the provincial forest fire management branch pirated the system, and after a government forester broke his back on the runway at Meadow Lake, turned away from testing it further. This chapter is the hardest to write , as you will see. Marilyn and I came out of this experience bankrupt.

Ω

There were two instigators. First, Bob Henderson, a former Oregon smokejumper (forest fire initial attack) and associate professor of forest fire science in the Faculty of Forestry at UBC. Second, I taught forest fire management at BCIT and was familiar with the difficulties of accessing lightning fires up remote valleys or high up on some hill in the Nimpkish Valley. Especially in 1961 during the summer shutdown. My dream was always to 'keep small fires small' using highly trained, fit initial attack crews. The primary skill needed was to burn out the fuel while using water as a backup only. Get rid of the forest fuel, and you get rid of the chain reaction combustion process at the development stage. Pouring high cost water on a fire does not remove the fuel. The corollary to prompt initial attack is prompt detection of the fire. Today, the protection of 'values at risk' in the form of subdivisions, Indian communities and recreational businesses have all added to the urgency of keeping small fires small.

Keeping fire out of the forest in the midterm does protect allowable cuts[21]. But when fuel loads build up and exceed stored energy loads that send flame fronts up into the forest canopy, we have expensive campaign fires as a result. Usually, we wait for nature to provide rain to severely retard a fire's behaviour. In the meantime, a few million dollars are spent.

A successful initial attack required a flame front not exceeding a flame height of 1.3 metres (an energy release rate of 285 watts per line metre).[22] Only prompt detection and rapid sky spider dispatch with the two-person crew lessened the chance of a fire escape. Above that level, the energy release rate from combustion was too high to fight a fire frontally. Government firefighters buy more time by bringing in air tankers to knock down the flame front. It then becomes necessary to use an indirect attack approach by pinching off the fire

118

head and attacking the sides first. This all increases cost. This approach needs the use of high cost water.

Bob's research in the Faculty of Forestry at UBC centred around the analysis of forest fire history covering the Upper Arrow Lakes north to Kinbasket Lake in the last 20 years. With his research assistant, Tanis, he classified the listed fires by size class and location. Tanis plotted the results on to topographic (NTS)[23] mapsheets. It was easy to show the largest fires occurred at higher elevations and up more remote creek drainages. He did this through an NSERC[24] grant (National Science and Engineering Research Canada.) The associated land base was within the boundaries of the Celgar Tree Farm Licence.

IFFS, 1971

In the winter of 1971, Bob and I developed a business plan containing expected work and forecast cashflows for a 'private enterprise' forest fire management service (International Forest Fire Systems Inc.). This plan broke down into three levels of work likelihood – maximum fire load, average fire load, and moderate fire load expected. These forecast fire load levels originated from that twenty years of Bob's research data. IFFS expected that various large companies, together with the BC Forest Service regions, would make up the clientele. Invoicing at $0.25 each allowable cut cubic foot became the basis for cashflow analysis using an average annual fire load. In practice, IFFS invoiced on an hourly rate for the crew, added in materials and supplies, and then included Okanagan's invoice for the helicopter. In 1973, the average invoice was $1,750. In comparison, a Forest Service fire exceeded $30,000.

IFFS registered federally and gained legal credentials in the spring of 1971. A contract through the West Coast forest industry followed. Until then, Forest Industries Flying Tankers (FIFT) supplied initial attack services. Its headquarters was on Sproat Lake, near Port Alberni. The IFFS firebase was at Campbell River. Equipment consisted of a Ford crewcab truck 4X4, chainsaw and hand tools. A federal unemployment subsidy covered payroll for the crew, three unemployed local men.

I stayed on the BCIT payroll, with the permission of the principal, Dean Goard. Bob and his researcher were covered by the research grant and were on the UBC payroll. The cashflow was limited to truck payments, subsidized payroll, food and accommodation, purchase of the best forest firefighting hand tools that money could buy, and fuel for the truck. IFFS only fought one fire that year, a lightning strike under a powerline in the Sayward vicinity. The use of air tankers on a fire alongside a powerline was dangerous.

Training and development was ongoing. IFFS used the Western Mines warehouse in Campbell River as a staging area. A simulated helicopter cabin in the rafters became the platform for trials. Bob and I developed the method associated with the delivery of two firefighters from a hovering helicopter plus their gear down a rope to a forest fire. (Ref. 'sky – genie© www.descentcontrolinc.com/ /[25])

The rope used was three hundred feet of woven nylon, ½ inch diameter, shock resistant. It was coiled into a 'birds nest' (donut with a hole) configuration It unraveled flawlessly as it was tossed to the ground from a hovering helicopter three hundred feet above the fire.

In the warehouse mockup cockpit the crewmen sat up in the rafters on a bucket, with the rope anchored to the rafter and carabiner[26] above them. Each wore a type B parachute harness with sky genie attached by another carabiner, and helmet. Bob played his role of a spotter. He had contacted air-sea rescue in Comox for help with a descent protocol and operations manual.

It transpired that the three local employees were heroin users big time. They learned the use of the hand tools and fireline construction. Each afternoon, the crew ran a long distance course overland in the Sayward forest. They wore their hair long. Bob ordered them to wear hairnets when working. These were essential as a safety measure. If hair got caught in the sky genie, the person on the other end of that hair could be scalped. The fuss that followed got IFFS headlined in the Campbell River newspaper. This is in the newspaper archives for sure. On one memorable occasion, I followed them up to a viewpoint looking across the Sayward landscape and arrived there just after a session with heroin.

IFFS had learned that two U.S. Air force crewmen rappelling out of both sides of a twin turbine helicopter in Alaska had collided as they descended. They fell to the ground and were killed. Large helicopters (Bell 204 or Bell 205) do create a stable platform in hover mode. These were a rare commodity in British Columbia in 1971. The decision was to frame the operations plan around the Bell 206B machine, single engine, cheaper to operate and way more available. There were 46 of these in BC in 1972, Vietnam War leftovers with one extra advantage – hardpoints on the fuselage above and below the exit doors used for machine gun mounts. Much of a Bell Jet Ranger's fuselage is fabricated from spongy aluminum to reduce weight. The engineers designed a trapezoid frame to serve as an attachment for the rappel rope. The crews referred to it as the wishbone (of a chicken). These hardpoints were placed ideally to do just this.

An aviation engineering company designed prototypes in Richmond, BC over the winter. They filed the final design for approval to Transport Canada. Also, during the winter to follow, Bob completed the operations manual for submission to Transport Canada using his National Scientific and Engineering Research Council of Canada funds. He included the approval for the wishbone design with the mounting protocol. Why? The federal operator's licence for any aircraft exactly states that no one can leave an aircraft in flight. Thus, by going down the rope from a hovering helicopter, the aircraft operator could lose his class 4 operator's licence. This was a hurdle to overcome. IFFS arranged with Okanagan Helicopters Ltd, (Carl Agar) based in Richmond to apply jointly for permission to void this need on rappel operations. It also applied to Transport Canada for the approval to activate the helicopter rappel operations system. An enduring relationship between IFFS and Okanagan Helicopters Ltd commenced.[27] The latter had satellite bases distributed throughout the province.

IFFS, 1972

Through the winter, Bob and I refined the business plan, and then approached banks for business loans. Canadian Imperial Bank of Commerce (CIBC) came through with a line of credit. Transport Canada issued an interim approval of the class 4 exemption and wishbone design. By late spring, IFFS continued with its plans for fire season 1972. Students from both UBC and BCIT were at the core of future events. A sky spider demo over the Seymour Armoury in Vancouver, and another on to the roof of the BC Forest Service ranger station in Kamloops, produced mild media coverage. The Company left the West Coast and set up its first permanent field base in Revelstoke in association with Okanagan Helicopters Ltd. The NSERC fire frequency analysis showed the upper Arrow Lakes region north to Kinbasket Lake was a key area for large remote lightning fires.

Figure 15 An A frame attachment with two sky genies on the rope before takeoff. Barry Manfield photo.

During the early part of that Revelstoke summer, Fred Baird, the base pilot, familiarized himself with the task of delivering an initial attack crew to a simulated fire. IFFS set up a scaffold at the Okanagan base and used this to train on. After two weeks on the tower, the trainees graduated to the helicopter. The approved operations plan mandated certification for the pilot, spotter and sky spider crew. Bob did the certification. Howard Murdock from the Mission tree farm trained everyone to fall timber safely. The success of this procedure was important. To descend the rope through the forest canopy was one thing, but the real work started with falling huge cedar trees that had caught fire.

Barry Manfield, CRSP, RFT, CSO.

Barry was a student at BCIT in 1972 and rapidly became a key part in the enterprise. He replied to a recent email:

"Yup, I started out in Revelstoke the first year. Ah... Sorcerer Creek, Carnes Creek, Mars Creek, LaForme Creek, Downie Creek, etc., etc., and the ever popular "Fish River" (formerly the Incomappleux River). I spent most of my time based out of Revelstoke in one role or another.

"Aerodynamics of the container"...? That's somewhat of an oxymoron isn't it, considering the "container" had the configuration of a large "brick"? The wax coated cardboard "Box" was developed during the winter 1971 / spring of 1972 and implemented that season. It was refined / improved the fall / winter of 1972. It was always a box while I was there, although I seem to remember someone saying that during the first year (before my time) a "canvas duffle bag" had been used. The "Box" idea may have been stolen from the US Forest Service, but I'd have to check back in my files to confirm that. "

Barry Manfield is the author of the photos displayed. He was a part of the IFFS 'can do' department. He currently owns and manages Sylva Management Services Ltd., Kamloops, BC. Barry and Mike Marcusson managed the Company's 'Stoneblaze' manufacturing division: harnesses, cargo nets, ditty bags, cruisers vests and helicopter shrouds. It was wonderful to get back in touch with 'Uncle Nick' after all these years.

The first sky spider sortie

It behooved the Company to seek improved initial attack opportunities. Fortunately, we moved the base to Revelstoke in the KOA campground in 1972, just before the passage of a lightning storm in the Columbia Valley up from the town. One day in early July Bob Henderson and the crew laid out all the rappelling gear in the Okanagan hanger in Revelstoke. Fred Baird, base manager and pilot, was out on fire patrol with Celgar's[28] woodlands manager following a lightning storm the previous evening. The Okanagan maintenance engineer was still in the building. He was familiar with the sky spider situation. He knew the protocol for bolting the wishbone to the hard points on the fuselage.

The woodlands manager was confused. He knew he had lightning strikes to put out, but the number of fires and their location confused him. Fred must have briefed him on IFFS during their fire patrol. When the manager saw the IFFS crew was ready to go, he directed Bob Henderson to send two crews. He provided a Celgar radio. Larry Huber and I had our fire under control by 1930 hours at the 5,500 feet level above Sorcerer Creek and so informed the woodlands manager over the radio. It was unheard of to bring a forest fire under control at that remote location under 3.5 hours with two people. He requested more crews right away. That was the start of IFFS' integration into regional forest fire management in Revelstoke.

Figure 16 Cargo pack – chainsaw on the extreme right. Barry Manfield photo.

By this time also, IFFS had designed, tested and built the cardboard cargo container. Each one held: a Stihl© chainsaw with 37-inch bar, pulaski, True Temper shovel, water for drinking, food and coffee. Earlier, though, I found myself flying at great heights over Revelstoke as we tried out various shapes of cargo containers. With the doors off, we noted the degree of spin, sway and stability as the cargo dangled below the helicopter.

Training

Two weeks of tower training removed the risk of aberrant behaviours on the rope. Those with previous mountaineering experience were the worst offenders because they felt they knew it all. Strength and endurance training continued all summer: a 6-minute mile, 50 pushups, 10 chin-ups on the high bar, as well as skipping and running in place.

Each of the two crewmen donned the type B parachute harness before entering the helicopter before a rappel. He tested his quick release for detaching himself from the rope once on the ground. He received a final inspection from the spotter, and then took his seat behind the spotter. Finally, the latter attached the rappel rope to the wishbone with a carabiner. He placed it between the feet of the first crewman to exit the aircraft once over the target fire. The first task was not to fall off the descent rope. As the helicopter lifted off, the engineer reached up and attached the fire tools box to the cargo hook under the belly. Meanwhile, the spotter attached a snub line to connect the dangling cargo box to the descent rope using a sky genie.

Figure 17 Sky spider descent training before fire season, 1973. Barry Manfield photo.

Note the carabiner holding him on to the rope and the ditty bag (personal effects) attached at his chest.

Once over the fire, usually at the 5,000 feet level or so,[29] the pilot put the aircraft into hover. He positioned it over the fire with advice from the spotter. This person carefully aimed the rope over the landing spot and let it go. Remember the wishbone A frame anchored the rope above the exit door. The crewman prepared his sky genie. He wrapped it around the rope the necessary number of times according to his weight. He slid the outer shell over the rope. He eased himself off the seat and on to the skid gently, the rope descending between his legs. If the crewman rocked the helicopter as he descended, he owed the pilot a case of beer. The crewman had up to 20 seconds to get to *terra firma* and disconnect himself from the rope using his 'quick release'. After the second crewman reached the ground, the pilot released the cargo using the live hook. The toolbox was now on the descent rope attached to its own carabiner. The crewmen controlled the rate of descent by pulling on the rope. A crewman could control his own descent by increasing the angle on the rope as this spiraled up through the sky genie. Remember, each sky spider wore pigskin gloves to protect his hands.

It was common for the helicopter engine to overheat in hover, especially, over 5,000 feet elevation in midsummer. The torque gauge would show red. That was why it was so important to be off that rope within 20 seconds. My record descent on a 300-feet rope was 17 seconds. However, I was not the fastest.

IFFS, 1973

Smokejumper operations
IFFS bought a Beech 18, twin engine aircraft for smokejumper operations in the Yukon and NWT. Mike Marcusson, ex U.S. Forest Service, joined the Company and trained a crew at Langley airport. The B.C Forest Service was dead set against employing smokejumpers in BC.

Figure 20 Graduating class, smokejumper school, Winthrop, Wn. spring 1976.

Bob Henderson in the back row, extreme left, with Larry Huber right beside him. IFFS trained with US Forest service crew that spring.

Figure 21 Dave Kyle piloting IFFS Beech 18 smokejumper aircraft. Barry Manfield photo.

Figure 18 IFFS CKT (Beech 18) looking for a spot to release smokejumpers in-training. Barry Manfield photo.

Training Okanagan helicopter pilots.

I went to Penticton to train Okanagan pilots in the delivery of IFFS crews to remote lightning fires. The pilots became familiar with the operations manual and descent procedures. Bob Henderson certified them.

Management and administration.

I borrowed $10,000 to buy a cookhouse and dining trailer and moved it to the KOA campground in Revelstoke. Friends, relatives and colleagues bought Company shares. A forest hydrologist from Rhode Island, U.S.A, Wil Poliquin, buddy of Henderson's became a shareholder and director. He was not around much due to consulting projects elsewhere. I earned my shares through the purchases I made on behalf of the Company. I developed forestry projects to provide cashflow during non-fire periods. Smokejumpers planted trees at Prince George on their way north. Crewmen based in Revelstoke undertook regeneration surveys, timber cruising and mapping, also slashburning. Both BCIT and UBC became sources for summer crews. Bob and I chose the best.

IFFS borrowed $45,000 through an RBC small business loan. Bob spent much of this on his new IFFS office. He located us in a big new warehouse complex in Richmond, BC, just across from runway 26 at YVR. He just had to acquire Steelcase furniture, a desk covered in leather, upholstered chairs, ash planking on the walls and a deep red carpet. On top of this, the IFFS summer payroll approached 160 workers, including management, pilots and field staff. It was sickening to watch. Bob's theory was that a host of wannabe clients would come racing in to his office. In fact, we did most of our business on the client's turf.

Although I gave up smoking on December 13, 1964, I did not abstain from cream and sugar in my coffee until June 13, 1973 in Revelstoke. I was so tired of opening the cargo pack on a fire, only to find that either the cream or the sugar was missing. I decided to forego these from then on. On one occasion, the coffee in the cargo pack turned into coffee beans. I put the beans in a sock and bashed it with the flat side of the pulaski. That was life in the fast lane.

Women crew

IFFS employed women on the forestry side, as well as on support to fire operations. The strongest organisations originate from a mix of both men and women. The women made sure the operations went smoothly. We recruited our women locally. They undertook forestry contract work all over. They were so easy to train. My one perplexity involved a woman's menstrual cycle and its possible attraction for bears. There was considerable *angst* within the crew at camp some times. When I could, I set up my tent close by. However, scanning the web in this age indicates that for black bears at least, menstrual blood is not an attractant. In my day we still worried though.

Figure 19 Revelstoke camp, 1974. Barry Manfield photo.

Barry Manfield and Mike Marcusson manufactured rappelling equipment, cargo nets, helicopter frost covers and timber cruising vests. They had used my old timber cruiser's vest as a pattern Then they added two unique additions. First they added an external pouch on the back. Second, my idea, they provided an increment borer pocket running lengthwise on one side. The vest was red. Mike opted for a distinct brand name 'Stoneblaze Manufacturing'. This Company had a reputation for quality goods made from quality materials.

In June, I arrived down the rope to a fire on a ridge above Arrow Lake across from Nakusp. Three big cedars burned hotly from a lightning strike. One of these had burned through at the base and had fallen over, spreading the fire outwards. We bucked up the fallen cedar and let it burn out overnight. We extinguished the fire in a second cedar with snow. The third cedar received the same treatment. The combination of snow and cold temperature overnight eliminated hotspots. The fireguard was cold. But that is not the end of this story.

The crew found a location for a helispot some distance from the fire. We felled trees and built a helipad prior to returning to base by helicopter. I then returned to the fire to rescue my pigskin rappel gloves from the base of a large cedar at the fire edge. The way back to the fire from the helispot was blocked by a nest of windfalls up to five feet high. I struggled through these and reached the large cedar tree. On looking up, I met a mother bear face-to-face. She was a cinnamon colour. I immediately scanned the surroundings for the presence of cubs. As she stood at full height and let out some unpleasant noises, I turned around and headed back to the helispot. A shot of adrenaline enabled me to jump across all the windfalls without touching one. One expensive set of rappel gloves written off. We boarded the helicopter with alacrity three hours later.

By this time also, IFFS mixed mud during a fire bust at the air tanker base at the Castlegar airport, in addition to its initial attack duties. The Forest Service used diammonium phosphate (DAP) as a fire retardant.[30]

The air tanker fraternity felt threatened by IFFS initial attack success. On one fire, high up on the shoulder of the mountain, an IFFS crew was completing the fireguard. They looked up to see a tanker load of DAP hurtling down towards them. What a mess, and how dangerous.

North Thompson fire

The Kamloops crew lived in the warehouse at the ranger station on Birch Island, up the North Thompson River. One afternoon, the fire patrol aircraft spotted a small fire way up a sidehill on a south-facing slope. The Forest Service delayed the dispatch of the IFFS crew until the fire was advancing up the sidehill. Once the flame heights reached the tree canopy, the crew was called in. Ugly behaviour, and dangerous. Elements within the Forest Service resented the presence of IFFS – 'turf' behaviour.

IFFS, 1975

Financing of operations came to a head in 1975. No fires out of IFFS, BC operations, period. Not one fire at the Revelstoke base was invoiced for in 1975. We based 1975 forecasted cashflow on an average fire season. No director was qualified to manage such a complex as IFFS with its severe falldown in cashflow. Consequently, we hired a professional accountant to set up an all-inclusive cost accounting system. His background was large international operations under the aegis of Peter Kiewit and Sons. He set up a system that was frightfully unfriendly. Add that to amateur ladies behind the admin desk in Richmond answering the phones (connections of Bob); the seeds of dissolution appeared. One recollects that many angry suppliers made life a living hell that summer.

During that non-fire year any rational manager should have laid off the whole crew, but these were students, and no one had the courage to get this done. Both Bob and I were aware of the difficulties of returning to school without pay cheques. We relied on a limited volume of forest survey work to supply cashflow, much of this earned by our women crew.

Bankers are a funny crowd. IFFS found that larger amounts of working capital needed processing further up the ladder. Decisions took ages to trickle back down. You might talk to a rep on the ground floor, surrounded by glass, develop mutually warm exhibitions of enthusiasm and earnest excitement, and perhaps are offered a coffee. However, by the time your submission hit the 20th floor, it was reduced to a stark array of bare numbers.

Thus, IFFS developed a strategy to obtain working capital/lines of credit. It subdivided its cost centre into a number of separate entities:

- Stoneblaze
- Smokejumper operations
- Forestry operations (Silviba Resources division)
- Rappelling operations
- U.S operations

The plan was to register these activity centres as subsidiary companies; although forestry was not set up separately until 1977. We got a bank loan in Portland, Oregon, but never set up an operation there.

Recollection is that 1975 depended heavily on forestry projects. The cashflow season was extended by taking on slashburning work, fall tree planting, timber cruising, and logging road engineering. One timber cruising job south of Revelstoke was completed just before Christmas, and in snow. I had just picked up the boys at Shelter Bay at the north end of Arrow Lake. The leased four-wheel drive Chev pickup then developed severe mechanical problems. The front drive axle spewed ball bearings and grease on to the road. We must have hitch hiked into Revelstoke and caught the bus for Vancouver. We abandoned all the camping gear and survey equipment.

Figure 24 IFFS (Bob) purchased this Aero Commander using Marilyn Richmond and Mrs. Poliquin as cosigners. Barry Manfield photo.

This is the aircraft that Bob Henderson flew back to Oregon after IFFS collapsed in 1978. He ran away.

Ω

I received my VFR pilot's licence and flew mostly in a leased Cessna 172, CG-HKP. I did get to fly the IFFS Piper Cherokee 180 CF-ROS once in a while.

Figure 20 Revelstoke airport. TR about to fly home to Vancouver, July 1975.

This enabled me to visit new bases at Cranbrook, Kamloops as well as Revelstoke. A three-week ground trip would now take three days or so. Add to this, management of tree planting operations in Prince George.

My flights into Cranbrook from the west were routed across Kootenay Lake, then over the hump into the St Mary's River valley. It leads right to the Cranbrook airport. This hump was a barrier between west and east. On one occasion, I looked down as I crossed the hump at 6,000 feet and noted a yellow Buffalo search plane traversing the landscape below. Several planes have flown right into that hump. It is especially difficult flying up and over this barrier on the return route from Cranbrook. The valley is narrow and does not leave much air room if having to turn back due to weather.

On June 15, I had a passenger with me on a return flight in the Cessna from Cranbrook to Revelstoke. Wil Poliquin. When I looked up into the St Mary's River valley, I noted that a big black thunderstorm obstructed my flight path. Wil's preference was to ignore it and keep flying up into the valley. A potentially lethal decision. After that day, I never entirely trusted Wil Poliquin. Instead, I followed the alternate route via Highway 3 to Creston, and then turned up into the south end of Kootenay Lake – lots of ceiling and air room.

135

Weyerhaeuser Vavenby

This may have been the summer when initial attack crews were based at Avola to meet Weyerhaeuser's initial attack needs up the North Thompson River. When not pre-occupied with forest fire matters, the crews were assigned other duties. One crew undertook log salvage on a cutblock only partially logged. Weyerhaeuser supplied a log skidder and diesel fuel. Another crew was given the job of recommissioning a large, disused bunkhouse-washroom complex parked not far from Highway 5 at Avola. The building was a mess. Itinerant folks travelling the highway had used the toilet facilities – no flushing water – and had flopped in the sleeping rooms. I took on the job of supervising a crew to bring life back to that bunkhouse complex in spite of imperious demands by a Weyerhaeuser woods supervisor. The woods supervisor had the crew cutting brush at blind corners on logging roads as a reward once the bunkhouse complex was rehabilitated.

Weyerhaeuser had turned over an abandoned house in Avola for IFFS use that summer. A pool of seeping sewage beside the front steps was one of the more memorable features of that accommodation.

IFFS, 1976

IFFS had fire crews at Kamloops, Revelstoke, Jasper, Cranbrook, as well as smokejumper and sky spider crews in the Yukon and NWT. Forestry projects took place out of Beaverdell, Midway, and Revelstoke and up on the Salmo-Creston sidehills not to forget the spring tree planting at Prince George before the crews went north.

Revelstoke dam-site preparation

That August, IFFS was contracted to dispose of the logging slash over a 16 hectare site at what is now the Revelstoke dam. The Okanagan helicopter pilot and I used a 'flying drip torch' to centre-fire the slash, then spread it out in circles as the fire sucked the smoke into the middle. The drip torch delivered jellied diesel on to the ground, ignited and in globs. That night, the Edmonton newspapers complained of smoke haze in that city. It was a fascinating experience to walk that huge column of fire and smoke across the logged area.

IFFS crews patrolled the fireguards to ensure that no sparks dropped into the adjacent timber. The forest ranger from the BC Forest Service in Revelstoke was on site also. His advice was hugely helpful.

And...

Crews mixed mud for helicopter bucket operations in Glacier National Park, Roseberry on Slocan Lake, and at the mouth of the Beaver River on the south end of Kinbasket Lake. There is always something immoral when fighting fire in a national park. It is supposed to exist as 'wilderness', but in practice, values at risk need fire protection, and these include scenery. One consideration has to be forest fuels management. Keep fire out of the park for too long, eventually you get a campaign fire that burns everything. Forest fire protection is about managing the potential energy budget. Energetic fire suppression provokes the storage of too much potential energy. Once a fire escapes the release of immense amounts of kinetic energy prevents fire suppression. The IFFS crew in the Park cut trail, renovated campsites and served as initial attack standby.

I commenced work fulltime with IFFS after teaching my last class at BCIT in June. Jim Dunlop, forest ranger at Lower Post became certified in rappelling operations and went down the rope to fight fire in that remote region. It seems that Henderson facilitated this as a means of favorable public relations with the BC Forest Service. I assumed that the Forest Service was invoiced for the rappel equipment required.

I had two more fires that season: one up the Incommapleux (Fish) River: [31] This was a fair distance southeast of Revelstoke. Its source was the south side of the Illecillewaet glacier. Fred Baird was piloting the Bell 204. He landed on a gravel bar right beside the fire – no need to rappel. It was in a jackpot of huge cedars and starting to burn vigorously. The three-man crew took out the chainsaw and started to fall the largest cedar I had ever seen. Those that weren't falling timber proceeded to cut an access path around the fire edge using hand tools. It was a hot day and all were sweating. Before the crew knew it, Fred had connected up his bucket and was dropping glacier-fed water on us. Boy, did that wake us up. Everyone was home in time for supper.

Figure 21 Bull fire site, waiting for pickup-Gary Weekes on left, TR in the middle. Who on the right? Barry Manfield photo.

The second fire was way up on the shoulder of Sorcerer Creek, at about the 5,800 feet level. It was in spruce-balsam and was going nowhere. The month was August. A low pressure system was moving in, and IFFS was called into Castlegar to mix mud. Fred Baird and Bob Henderson in the 206 came to pick up the crew – no helispot – just an airlift on the end of the rope up through the canopy. I made sure the other guy's harness was secure and attached him to the end of the rope. Fred extracted the first man, and then swung him down to the gravel bar in the creek. He returned for the cargo box. I clipped it on, and then waited for my turn to be plucked up. In a while I heard the helicopter climbing back up to the fire site. The rope came down; I attached myself and was lifted up and out. Then the fun started. An unstable low pressure cell had moved into the valley and up over the fire area. I started to oscillate at the end of the 300-feet rope at 5,000 feet plus above the creek. I travelled in a wide arc under the helicopter. Fred could not move forward since he was fighting to preserve stability in hover. While I swung, the rope ran up against the bear paw[32] found on the end of the skid. Each time this happened, the noisy 'twang' of a stretched rope travelled all the way down to me. I passed over one patch of windfalls on the sidehill seven times before Fred could lose altitude and gain the gravel bar below. All the while, Bob was using his foot to keep the bear paw from cutting the rope.

Bear paws are attached to the skids and are used to spread the machine weight on softer ground. Once back on the gravel bar below, we carried on to Castlegar to mix mud. Then it started to rain!

After fire season, the crew slashburned (broadcast burns), as well as burning slash piles on Weyerhaeuser landings.

Figure 22: Randy Lafferty, TR, Phil Cloward, Larry Huber, Mike Marcusson, young buddy. Barry Manfield photo.

That fall, I had crews on tree-thinning projects up Harrison Lake. I used a wide logging road designed as a runway for light aircraft. When I taxied for takeoff on January 9, 1977 for Vancouver YVR, the nose wheel fell into frost pocket. The prop hit the ground. Rather than take off, I parked the machine (ROS) and got home by other means. By this time I had logged 278 VFR[33] hours. This was the last time I flew in ROS. A mechanic flew out later, checked over the plane and flew it back to YVR. I did not fly as a pilot ever again. Could not afford it.

Silviba, 1977
My management duties took me away from direct involvement with initial attack. Over winter I developed a business plan to gross $171,000 over the next year. Silviba received some large juvenile spacing jobs on northern Vancouver Island in September. Beforehand, I visited a lawyer in Richmond and signed all the paperwork

subsequent to the registration of Silviba Resources Ltd, pursuant to the Companies Act. Bob Henderson's director's signature was still needed though. Since I had to leave town immediately, I relied on Bob to complete the registration. By this time, Silviba had a Workers Compensation board registration number and a bank account at the main branch of the Bank of Montreal on Burrard St. in Vancouver.

Silviba had two tree-thinning contracts: one with Canfor in the Nimpkish, and a second contract west of Port McNeill with Rayonier Canada. The crew of 35 camped and completed both jobs before returning to Vancouver in November. Fire caused by an overheated stove pipe destroyed one tent. Some 40 chainsaws needed overhauling. But, the crew could barely scrape up enough money for gas and the ferry from Nanaimo into Vancouver. I went immediately to the offices of both Rayonier and Canfor downtown. It did not take long to be told that Revenue Canada had scooped $143,000 owing to Silviba just three hours before I got there.

Shortly, I determined that Henderson had not signed the Silviba incorporation papers, and thus the revenue in question belonged to IFFS. This forced me into personal bankruptcy. Marilyn and I lost two houses in the Dunbar district on Vancouver's west side to the bank.

Shortly before I was to leave for Prince George to join Industrial Forestry Service Ltd, I went to Victoria to visit with Don Owen, in charge of forest fire management for the BC government. I urged him to ensure that the IFFS rappel procedures would not be lost if IFFS went down the tube.

I heard that Henderson flew the Aero Commander into the U.S. sometime in 1978, and was never heard from again. Bob was addicted to Roots footwear, by the way.

Ω

Chapter 12:

Industrial Forestry Service Ltd (IFS)

1978

Indirectly, my connection with IFS went back to 1964. I was presented with my certificate of professional forester registration (No. 403) by Larry de Grace at the annual meeting of the Association of BC Professional Foresters in February of that year. It was held in Victoria at the Empress Hotel. Larry also signed the certificate in his role of president. Larry started IFS. He later died while swimming at the pool in Prince George. When I arrived to start my senior forester's duties in early January 1978, Harry Gairns was president, together with Al Nevison and George Kondor. These formed the executive committee. The office was located at 1595 Fifth Avenue Prince George, BC. V2L 3L9.

I could not have done better than to join this group of highly effective and efficient professionals. Harry Gairns, RPF, P.Eng was an unassuming forest engineer with a mind as sharp as a tack. The organisation was flat: four cells of employees, each headed up by a senior forester and a number of junior foresters, timber cruisers, engineering assistants and compassmen. When I arrived, the Company shuffled everyone around, so I formed another cell. And I was given a similar core team that peaked at 16 men. My forester-in-training and senior man in the cell was Bruce Barry.

My first job was to join my team on a timber cruise in the Narcosli forest. My first ever snowshoe experience. It was a warm January week, on 5 feet of wet snow – barely a crust. I ran compass that week and was able to observe IFS procedures from low down on the totem pole. After work the gang hightailed it back to the vehicle leaving me to drag my worn out, saturated self to follow in the rear. It was dark, the moonlight threw shadows on to the snowshoe tracks, and I was barely able to stay on the path. Just before arriving back at the vehicle I had to traverse a steep bank on snowshoes that seemed to go up forever. One never forgets an experience like that. However, it was all part of a typical day's work. That crew taught me much in a very short time.

While at IFS, I completed two long-term timber supply studies. This was in the days before sophisticated computer modelling, and I determined the allowable cut for the next 200 years on a sustainable basis. However, I got the job done in both cases. The first was for Canfor in Prince George. I used all the landbase within its area of interest. My crew surveyed cutplan locations from Burns Lake to the west and to McBride on the east side of Prince George. To the north, it surveyed the entire west side of Williston Lake, all the way up into the Parsnip River. At times the crew reached as high as 16 members. The clients included three pulpmills and seven sawmills. My favorite customer was Bob Goodwin, who owned Fort St. James Lumber. This was probably due to his need for more of my involvement in the timber supply logistics; unlike the large firms that employed entire departments to plan their forestry operations. The second involved Finlay Forest Industries, a pulpmill/sawmill plant at Mackenzie. Here, the wood supply analysis involved the use of timber cruise plots to adjust Forest Service volumes for its operating area. Because of Forest Service sampling protocols, the volume available as timber supply was significantly more conservative than experienced in practice.

The Forest Fire Simulator (2)
My one retrieval from IFFS before I walked away forever was the portable fire simulator. It was contained in a fiberglass case that closely resembled a white coffin. When set up, it performed identically to the fixed facility at BCIT. IFS allowed me to continue my industrial forest fire management training using the simulator. It travelled well, either in the back of a pickup or in a small aircraft. One of my last three-day training sessions occurred at Masset in a Macmillan Bloedel logging camp, on the very north end of Graham

142

Island on Haida Gwaii. I had dreaded having to undertake forest fire management training in the Charlottes. It always rained there, so who worried about forest fires?

Well, this was the summer of 1980, and the Charlottes were experiencing a high drought code. Given the path of logging slash from Masset all the way south to Queen Charlotte City, all Graham Island could have burned that summer. Talk about a long fuse just waiting to be lit.

The Macmillan Bloedel camp included married quarters. The presence of these automatically means a bulldozed ribbon of old slash surrounding the camp and logs pushed up to make room for streets and houses. Married quarters also contain kids and kids light fires when they are bored. This is inevitable, totally predictable. That afternoon wind blowing strongly from North to South was cause for great concern. I simulated three fires at Masset:

- The first in a slash pile adjacent to the mechanics shop in camp (during working hours)
- A night shift welding fire at a logging machine 30 km from camp
- A lightning fire on a weekend 25 km from camp.

'God' can get away with a lot.

BC Universities Commission

Harry Gairns sat on the BC Universities Commission. IFS contracted to investigate the relationship of technical forestry training at institutions such as BCIT with the professional forestry degree program at UBC. Harry Gairns was the mentor, and I did the leg work. Since I had teaching experience both at BCIT and the graduate studies program at UBC, I was familiar with all the movers and shakers. The essence of the questions to be answered was:

- How much duplication occurs in course material between UBC and BCIT?
- What credits may a tech student from BCIT bring as he or she enrolls in the Faculty of Forestry in order to complete a professional degree?

I had forecasted the obfuscation within the Faculty of Forestry. 'Turf' is hard to remove, but I also knew that younger faculty would support a methodology for a transfer of credits between the two institutions. For one thing, BCIT forestry grads after the 2-year program were hired immediately, whereas UBC grads were not only offered less salary to start with, but there were also fewer job offers available to them. BCIT graduates were like kangaroos – they came out jumping.

Conclusions and recommendations from both institutions resulted in a format that gave aspiring professional degree students from forestry programs at BCIT and other institutes almost a year's worth of credit.

Lastly

In the 3½ years at IFS, I sponsored four of my crew as they earned their professional forester credentials. On one smoky spring day, Harry Gairns allowed me to ride up to Fort St John in a forest nursery truck full of seedlings. My destination was a Forest Service fire management team at the junction of the Beatton River and the railway. The fire was threatening the railway bridge over the River. The management team was from Nelson.

As soon as I got there, I walked into the office portable and was treated as if I had just escaped from prison. At least they gave me a radio. I was put at the head end of the fire and supervised a D9 cat plowing fireline. I had no idea where I was located. The flames were at tree height and I lost sight of the cat. I felt like a complete idiot, and was worried out of my mind for the operator. A Conair air tanker zoomed overhead with branches sticking out from the tail aileron. Somehow, I finally found the cat; the operator had trailed back in behind the flame front.

The next day I was given two helpers. Our job was to patrol the fireline on the back end of the fire and dig out any smoldering hotspots. This was my first fire on permafrost – unique.

On the third day, they gave me back the cat. It was a D9 – lots of torque. I sat beside the operator as we crashed through trees at the fire edge. I was almost killed when a tree dislodged the spark arrestor, weighing 3 kg, from the exhaust pipe. It was similar to a catapult moment. The tree, under tension, let loose, hit the spark arrestor on the rebound, which then flew like a rocket into the back of the padded seat between the operator and me.

The permafrost rapidly turned into mud behind us. By the end of the day we had completed a guard around the fire.

And....

On the last working day each year I was the last one in the office. I spent that quiet time staring out of my window at the parking lot below. Why? I was running over in my mind the anticipated workload for the year to come. I did this at the end of 1978, 1979 and 1980.

I decided at the end of 1980, as I looked out of the window, that the forest industry was on the verge of, or had reached, a definitive downturn in the forest products market. The signs were Canada wide. This got me thinking about moving on. I did not want to be a drain on the Company's cashflow during a slow revenue period. Consequently, when offered the job of executive director of forestry by the Saskatchewan government in September 1981, I took it.

Ω

PART TWO
Saskatchewan

Chapter 13:

Allan Blakeney was still premier

The hiring process, 1981 fall

I first read about the job of 'executive director, forestry' posted by the government of Saskatchewan in a professional journal while still working in Prince George in the summer of 1981. I did not look at it twice. First, no one in my circle thought there was any timber in Saskatchewan. Second, I was busy with locating, then surveying in a forest systems road in the Finlay River, north of Prince George at the top end of Williston Lake. It was a fly-in job by helicopter.

In July, I received a phone call from a recruiter in Vancouver, who sent me an air ticket and a request for an interview. At least, I thought, it would serve as a free trip to visit my parents in Port Coquitlam. After lunch with the recruiter, he left me alone to study several documents touching on forestry in Saskatchewan. These included copies of reports from forestry consultants. My conclusion, after an afternoon in the boardroom, was that Saskatchewan forestry documents left a lot unsaid. In addition, the consultants' reports, though technical, lacked measurable terms of reference and were incoherent. I returned to Prince George feeling confused, but willing to explore further.

The Saskatchewan government arranged to interview me in Regina. I flew PWA from Prince George with stops in Dawson Creek and Edmonton. As I left from Edmonton, the sun was setting behind me. The night turned into a blue-black universe, pure velvet, interspersed with twinkling lights here and there, as I flew over the

Prairie. Then, over my shoulder, the last streak of sunset. It impelled an unforgettable feeling.

The next day, dressed in my grey flannels and blazer, I showed up for the interview in a government boardroom, and found there was a two-hour delay. It was a late August day in Regina, brilliant sun, 30 deg. C and a small breeze – combine weather. I walked around Wascana and returned at the appropriate time.

Francis J. Bogdasavich was the deputy minister for Saskatchewan Tourism and Renewable Resources. He chaired the interview, flanked by top civil servants in Finance and perhaps others. Since I had no anxieties about the result of the interview – take it or leave it – I spent most of it lecturing to those assembled. This included the need for an acceptable forest inventory that could be 'taken to the bank' if necessary. The next day, I flew back into the Finlay River and rejoined my crew.

Two weeks later, the helicopter pilot relayed a message from my wife: 'return to Regina for a meeting with the minister'. On my way back to Prince George from camp, I heard over the pilot's radio there was a fire on Punchaw Crescent. My family lived at 4293 Punchaw Crescent. I became very anxious. The pilot let me use the phone in Mackenzie, the pilot's base. I phoned home, and my wife told me the house involved was a neighbour down the street, and that I had to contact Regina.

Soon after that, I was back in Regina at the legislature for an interview with Honourable Reg Gross. Reg was in the basement at the very east end. Lots of sun coming through the arc-shaped windows high on the wall. It lit up Reg's Royal Proclamation appointment scroll – very impressive. It was immediately obvious that Honourable Reg Gross knew 'sweet tweet' about anything to do with forestry. What little he did know, he had learned from a German neighbour. Anyway, he had to get back to his combining. After this visit, I knew that Saskatchewan needed me. Frank Bogdasavich offered me the job on the steps of the Legislature as we left. He accompanied this with a strict instruction to stay away from the government forest fire protection branch. Little did I know that my appointment was an 'order-in-council' (OC) appointment, thus one who serves at the pleasure of government? Should I have severely dislocated my family from its comfort zone based on an unsecured position? The answer should be 'no'. However, the letter of offer mentioned a 6-month probation period. This would not have been a condition of employment if one merely served 'at the pleasure of government'. My salary was set at $51,000 annually.

I returned to my work up the Finlay River and waited for someone to announce a starting date. I heard later that conspiracies behind the scenes were intense. In effect, I was to take over from John Burton who appeared to have an associate deputy minister's rank. But John would not go. Internal strife within the senior NDP hierarchy delayed my appointment. This delayed the preparation of Allan Blakeney's political platform before the next provincial election in May. Unknown to me, forestry was to be a big part of that platform. However, they had left it too late.

Finally, I arrived in Regina on Thanksgiving Day, 1981. The family stayed back home to finish the school year. I found a basement suite on Queen St., and walked to work at 3211 Albert St. most days. When I had finally arrived at work for the first time, the whole east side of the fourth floor was open – all dividers stacked against one wall and all telephones disconnected. This is what it took to get John Burton out of there. Thus, my first desk was that formerly used by John in that final period. I found it wedged into the extreme southeast corner of the fourth floor, and I lacked a telephone and a secretary.

When I finally connected with filing cabinets, an office, secretary (Lucille Bitz), and walls, I spent the weekends examining a filing cabinet full of consultants' reports. Michael Decter had had a hand in preparing reports and associated analyses. Even here, there was no thread that leads to conclusions about industrial capacity. Also, no discussion on Crown revenues produced by forest operations, or anything to do with forest sustainability.

"Getting to know you"

Frank Bogdasavich was an excellent boss. He ranked up there with Roy Jewesson and Owen Hennigar from my Canfor days. He paraded me in front of some of the movers and shakers in Blakeney's cabinet, as well as key civil servants. He was especially cautious

151

when introducing me to one minister, Ted Bowerman, a former smokejumper, who later retired to Shellbrook. Frank was Allan Blakeney's troubleshooter.

Frank introduced me to his staff in Prince Albert at a special lunch in the Marlboro. All the key folk were foresters: Walter Bailey, Forestry Branch director, Murray Little, manager of the forest management wing, Steve Price, silviculture and Jamie Benson. The latter was manager of forest inventory. I received a controversial surprise: there was not one, but two Forestry Branches – the second entrenched within the Department of Northern Saskatchewan (DNS) out of La Ronge.

My 'terms of reference' included an informal instruction from Budget Bureau to fire key Forestry Branch managers, most particularly, the manager responsible for forest inventory. This official had warned government that the Crown sawmill in Carrot River had insufficient timber supplies to keep operating. This was in 1976. On the way back to his office from Carrot River, with all the difficult numbers in his briefcase, he heard on the news the government was not going to shut down Carrot River sawmill. Another matter, in 1980, also involved another relationship between Forestry Branch and Treasury Board. Canfor from Vancouver negotiated for the purchase of certain Crown forestry assets, mainly the pulpmill in Prince Albert, from the Saskatchewan government. This prompted a 'good guy-bad guy' strategy. The good guy represented the minister responsible for forestry at one end of the Legislative building, and the Department of Finance, the bad guys, ensconced themselves at the opposite end. Canfor needed comfort from: a) a fair price for the assets; b) an assessment of the quality and quantity of the provincial forest inventory and c) a reasonable decision on the level of Crown dues. Canfor became so frustrated at the games going on within the government negotiating team that it removed itself and returned to Vancouver (confidential source).

I did not fire anyone. Transfers and retirements solved the matter of duplicate staff and other irritants. Mostly, I shuffled the managers around to create a more coherent management team. How can you fire someone you have barely met? The core of unhappiness created by the loss of comfort zones was never addressed until after I was fired in 1984. Then the managers slipped back into their former positions.

Later that fall, I joined a party at Frank Hart's house on South Albert St. I was to meet Allan Blakeney, the premier. Mr. Blakeney brought up matters such as plywood, but could not explain anything specific.

On October 18, Mr. Bogdasavich asked me to sit in on a Crown Investments board meeting at the top of the Potash tower in Saskatoon. The agenda was mostly to do with the pulpmill and wood supply. The woodlands manager for Prince Albert Pulp Co. (PAPCO), an orphan Crown corporation, reviewed forestry operations within his jurisdiction. All the ministers around a huge boardroom table lounged in various stages of undress, some with shoes off. The chair called a halt for supper. I found myself sitting at a large table in a restaurant on 8th street. The atmosphere became more intimate as supper progressed. There was a general discussion involving the introduction of colour TV in the Leg, much leg pulling and laughter. One minister, Honourable Don Cody, wore a rich brown suit, orange shirt and yellow tie. I remarked to him that his colour scheme may appear too rich for wider audiences. (Just joking). I felt humbled by all that collective power sitting around that table.

As I got up to leave, the electricity went off in Saskatchewan. Luckily, a taxicab outside the restaurant drove me to my suburban. The city was in a total blackout – streetlights, traffic lights and highway signs. I caught myself driving up the highway to Blaine Lake before turning around and rerouting to Prince Albert. As I approached Prince Albert the lights came back on. I found out later that only one light bulb burned bright in Saskatchewan that evening – in the service station washroom in Creighton. Thanks to Manitoba Hydro? Sometime later, Don Cody became the mayor of Prince Albert.

By Christmas 1981, I had visited:

- The woodlands staff at the Crown Prince Albert Pulp Co. Ltd. (PAPCO)
- The Crown sawmill at Big River

- The Crown power pole and fencepost treatment plant in Prince Albert
- Crown sawmill at Carrot River
- The Crown plywood mill at Hudson Bay (Saskply)
- The oriented strandboard (OSB) plant operated by MacMillan Bloedel at Hudson Bay. It was the first one ever built (1965)
- The Simpson Timber sawmill at Hudson Bay
- The Crown sawmill at Meadow Lake
- Crown furniture factory, also in Meadow Lake.

What was not on my agenda were 143 small sawmills and fencepost treatment plants listed in the Saskatchewan forest industry directory compiled in 1979 by DREE (Canada Department of Regional Economic Expansion).

1982

Budget Bureau gave me until January 4 to get my Forestry Branch budget into the system. By cutting back on a trip home for Christmas, I not only completed the budget, but also increased the tree planting allotment by 1.5 million seedlings for 1982. And it was approved. I assembled a team of mixed abilities and backgrounds to update a forest policy for the province. This included an economic rationale for the role of forestry, as opposed to agriculture, potash and uranium. I inherited an economist from the World Bank. He turned out to be totally inarticulate in the discipline of forest economics. His replacement was Tony Baumgartner, a recent commerce grad from the U of S hired by the team as a researcher. Another team member, Garth Whyte, extracted from a public policy development unit somewhere in government, rounded out the group's quotient of knowledge, skills and ability.

Forestry Branch was headquartered in Prince Albert on 12th St. East, in the building SaskTel later refurbished for their regional office. Here, I worked with Andrea Atkinson, a management forester in the Branch, to calculate allowable annual cuts for the commercial tree species of Saskatchewan. She did this by region: Hudson Bay, Prince Albert, Meadow Lake and La Ronge. In those days, calculators were primitive, but she got the job done. The forest growth rate minus expected losses from fire, disease and insects becomes the net allowable cut available. This, in turn, becomes the Harvest Volume Schedule. See Saskatchewan government 2012 report on Saskatchewan's forests.

Dr. Tony Kozak was a professor of forest measurements at UBC. Forestry Branch hired him to produce taper tables for the key forest ecotypes in Saskatchewan. This information, in turn, told management foresters the proportions of pulpwood, sawlogs and peelers in each stand of timber. Remember, the provincial Crowns own 93% of the forestlands in Canada, which is why Forestry Branch needed this information. The report showed the provincial Harvest Volume Schedule (sustainable yield) was almost double the timber harvested. There was (is) room for an expansion of the forest industry in Saskatchewan.

Following this, Forestry Branch hired Industrial Forestry Service Ltd of Prince George (my former colleagues) to study Crown timber dues in various provinces across Canada. So, the Saskatchewan legislation detailing Crown charges was amended to reflect the recommended outcome for Saskatchewan.

The government changes

The government changed in May 1982. I worked for a new deputy minister, Bill Klassen, and a new minister, Honourable George Mcleod. The Conservatives had sat on the provincial sidelines since 1935, and entered government like gangbusters. The knives came out, and there was bloodlust throughout 'the kingdom'. One Friday, after quitting time, a coonskin coat walked into my office and sought a forest nursery brochure. I had never seen one. I found it eventually. There was a picture of Reg Gross on the back cover. This was probably incorrect political behaviour. But, so be it.

Bill and George had summer cottages side by each in the Bronson forest in the Meadow Lake vicinity. Bill started his career as a school teacher in a one-room school at Abernathy. I think George was a former principal, or had taught high school in Meadow Lake. Anyhow, they were buddies. In government, the heads began to roll. Some folk hounded by new deputy ministers suffered heart attacks, or were leaned on until they quit. Frank Bogdasavich was fired, but did not see this coming. He was devastated. An ex lumber scaler in the government economic development arena in Regina, who I got to know because of his job, was hounded almost to death by his new deputy minister. One government minister, Honourable Eric Bernston, future Senator, was in charge of the witch hunt. Since 'death hath no fury than a woman scorned' – the Conservatives last held government in 1935 – these folks were out for blood. Yes, there were NDP carpetbaggers to get rid of. However, amongst the carnage, went some exceedingly bright professionals, such as one woman high up in Finance, and Frank. That is one downfall from witch hunting, you throw the good out with the bad. The lady from Finance was snapped up immediately by someone similar at the World Bank. You would think that tourists were nonpolitical by nature, but Frank Hart, director of tourism in my department, was also terminated. Today, Frank is president of Greystone Management in Regina after serving as a senior executive in Crown Investments Corporation, if not its president. You do not get much more senior than that on the Prairie. Sometimes, everything works out for the best.

Bill Klassen and I had a professional relationship. I told him what I was hired to do in Saskatchewan, and I would keep doing this until someone told me to stop. I felt far more comfortable around Honourable George Mcleod. That summer, Bill and I joined a picnic in New Brunswick to set the agenda and itinerary for the annual CCFM[34] meeting at Cornerbrook, Newfoundland later that fall.

"There has been a long tradition of cooperation between the federal and provincial/territorial governments in forestry matters." According to one federal government website.

Bill and I arrived in Fredericton on a Sunday evening sometime in July. We joined the other deputy ministers and chief foresters from across Canada the next morning to develop an agenda for the upcoming CCFM meeting in Cornerbrook. This lasted less than hour. We got on to buses and proceeded to become the guests of the New Brunswick government for the rest of the week. It was a continuous party. Obligatory trips covered an Irvine site prep (preparation of the ground prior to tree planting) using a crushing machine, also a plantation or two.

Of interest, we passed through the Acadian settlement area, toured an old fort, *about 1650*. Before that, we had spent a day and a night fishing on the Restigouche River. Unknown to us, though, the Indians had blockaded the mouth of the River. No bites. No fish. Great party.

We spent the last evening as guests of the Maritime Lumber Bureau in Caraquet. We feasted on lobsters – fresh. I had three. The first I had ever eaten. And the last until I ever return there. Bill Klassen was a hit. He brought along his guitar and had the whole bus, including the driver, singing: *'Goosed by a Moose in the Spruce.'*

I ended that trip hugely depressed. I had never experienced such a terrific waste of taxpayer money. However, the saga continued.

Later, in the fall, I accompanied both Honourable George McLeod and Bill Klassen on a flight from Saskatoon to Cornerbrook for the annual CCFM ministers' meeting. George was drunk, having attended a provincial tourism association meeting in Saskatoon beforehand. We arrived in Cornerbrook, had supper, and then hung around in the lounge area of the hotel. Bill had his guitar. By 1.30 am, there was no one left to party with – just Bill and George. I had just got to sleep when there was a knock on my door. Bill had no one to drink with and needed company. I kept quiet and Bill, upset at the lack of response, shuffled away.

After a token meeting the next morning, the entire delegation found itself spread among several boats in the harbour. I found myself on a small tug with other like-minded delegates. We began to fish and caught a number of cod. Later, we were boarded by two partying ministers, drunk and disorderly, Honourable George McLeod and his Newfie counterpart. After clambering aboard, they threw all our fish, now dead, back into the water. What a waste.

I went home with two bottles of Newfie Screech. I found out that this was less expensive in Regina. My family moved to Regina that summer and negotiated a mortgage at 12 percent interest. The Devine government had bought the election with the provision of government subsidies to offset brutal mortgage interest rates.

At some stage, I took part in the annual Department meeting held at Cypress Hills Provincial Park for the directors of the various branches. There were about twenty of us. All I remember is one session lasting well into the night time with a few hostile directors. As the drinking progressed, they became even more upset with Forestry Branch behaviours. The hostility baffled me, especially since Bill Klassen was 'priming the pump'.

The names of three senior folk are retained in my mind: Lyle Lenson, ADM administration, Barry Tether, director for department operations in the south, and the regional director at Meadow Lake. I had much respect for them as colleagues.

Once again, Budget Bureau informed me that I had until January 4th or so to have my budget submitted, and once more, I surrendered Christmas time for paperwork.

Ω

Chapter 14:

Further struggles to build a provincial forest policy platform, 1983-84

1983

Forestry Branch became forestry division with two branches: Forest Management Branch under Walter Bailey, and Forest Operations Branch under Frank Flavelle. In Regina, the policy development group was gelling; it worked with improved information for the provincial allowable volumes by species and region. Classifying the provincial forest inventory included log quality: sawlogs, pulp and peeler grade material. Thus, it was possible to assess industrial capacity against available timber supply.

Early in the year, Bill Klassen became very ill, and stayed in the Royal University hospital in Saskatoon for an extended period. As the forest policy development process took shape, I dealt directly with Honourable George McLeod. However, I dropped in to see Bill to keep him abreast of progress. Then, George was transferred to another department, and Honourable Neal Hardy took his place. However, before George left, I gave him a draft forest policy framework document. George put it into his jacket pocket and promised to take it to cabinet for discussion. Neal was a down-to-earth former union leader and railway worker in Hudson Bay. He ran a grocery store and fuel pumps on the highway leading south from the community. He asked for periodic updates, as well as providing feedback, for the policy development process. I hitched a ride up to Prince Albert on the government jet one afternoon. Neal and his executive assistant, a young accountant, also from Hudson Bay, started to discuss a new model of air tanker that was powered by two jet engines. I told them you could not drop retardant on to a forest fire with jet-propelled airtankers because jet engines would be unable to slow the speed down sufficiently. Well, I got egg on my face. Seemingly, Saskatchewan was pondering the lease of new air tankers, powered by jet engines.

By the spring, I had a third minister to contend with, Honourable Bob Pickering, who was a former curling champion and who lived in Bengough. He seemed hostile. Bill Klassen was back full-time. National Forest Week was proclaimed for a day in May. I was instructed to round up some white spruce seedlings for Honourable Bob Pickering to plant in a ceremony on the grounds of the Legislature. When that time came, he failed to show up. I planted them on the front lawn of our house at 56 Richardson Crescent in Regina's south end. They are doing well and are now about 34 years old. I suspected that trees did not play much of a role at Bengough, SK.

Forestry Branch faced two hurdles that summer. First, we started to negotiate a new Canada-Saskatchewan forest resource development agreement known as FRDA. Second, we began to negotiate an area-based, sustainable long-term forest management agreement with the Crown pulpmill. The FRDA covered continuing federal forest research projects (mostly long-term forest growth studies). Another section involved private woodlot management plans. The third section encouraged economic development – mostly aimed at Indian forest management opportunities (on-reserve). The template was formal and left little to the imagination.

The pulpmill (PAPCO) negotiation was complex and invidious. It took place in Forestry Branch's basement boardroom in Prince Albert. Since it was a Crown corporation, the owners representing PAPCO sat on one side of the table in the am, while Forestry Branch sat on the other side. Negotiators armed themselves with all the prerequisites for a long-run forest management model.

After lunch, the same individuals, now representing Treasury Board, tried to demolish all that progress made before lunch. The big argument centred on the boundaries of the proposed forest management licence. In the morning, the "owners" realised that the larger the area within the proposed boundaries, the larger the allowable cut. So, the "owners" were happy to include the landbase north to La Ronge, then over to Beauval on the west side, and on the east side to Smeaton. In the afternoon, Treasury Board (the same "owners") tried to reduce average hauling costs by stating that the distant timber was uneconomical to include. They wanted to redraw the boundaries. This started the demise of one TR, as you will see.

The Prince Albert Pulp Company (PAPCO) query

On one afternoon that summer, in my Prince Albert office, I received a phone call from a business development specialist in Vancouver. He was seeking fresh investment opportunities for a major client. This person was also an adjunct professor in forest economics at UBC. I had met him previously, and considered him 'real'. The essence of the phone conversation to follow was this:

On the down side, I was critical of PAPCO's delivered wood cost, twice the industry average. The pulpmill relied on open market pulp prices that, in 1983, were in the basement. Canfor had tried to buy the facility in 1980, but failed. There was no external wood chip supply, since the pulpmill management had a long-standing fight with the two Crown sawmills over respective wood supply territories. PAPCO was bypassing pulpwood stands to get at the sawmill sawlog supplies. From a public forest management perspective, this downfall in economic gain appeared negligent. Last, some sawlogs delivered to the pulpmill yard – an oxymoron? – for chipping into woodchips were too large for the throats of the woodchipping units. PAPCO took these 'oversize' logs to a spot behind the pulpmill and buried them in an effort to conceal the 'crime'. However, these caught fire at awkward moments from spontaneous combustion. This was a constant burden for the Prince Albert Fire Department. There was a desperate need to integrate the sawmill and pulpmill wood supplies.

On the upside, PAPCO had received a grant from DREE (Department of Regional Economic Expansion), a federal program, now expired. With these funds, the pulpmill developed systems for pulping hardwoods. A batch of hardwood logs had arrived in Sweden for processing. I had inherited a credenza on my arrival. This was full of pulp samples. The results suggested that a 70:30 ratio between softwood and hardwood would make good pulp. This meant that Forestry Branch could double the provincial commercial wood supply, given that hardwoods are fifty percent of the forest available. Finally, I briefed the enquirer on the high quality of the labour force. Since many of the pulpmill workers were part-time farmers they knew how to work.

To return to the PAPCO/Saskatchewan Treasury Board negotiations with Forestry Branch: I referred to the 'Weyerhaeuser model' as desirable in negotiating a forest management agreement (FMA). This would direct the forest resource to the highest economic gain: pulpwood to the pulpmill, sawlogs to the sawmills and the highest grade of white spruce (peelers) to the Saskply plywood mill in Hudson Bay. See '1986' for further developments on this topic.

Forest operations unit and small operators

In early 1983, Frank Flavelle retired as manager of forest operations. I missed him. He was a professional and understood his role. He had avoided the traps of mediocre forest legislation and timber regulations you could drive a log truck through. Fortunately, another operations forester, Vic Begrand, replaced him. Vic was formerly an industrial forest manager with MacMillan Bloedel at Hudson Bay. The forest operations unit had resource officers based in the various district offices: Meadow Lake, La Ronge, Prince Albert and Hudson Bay. Though these were an ambiguous crew, beset with unceasing small operator problems with wood supply.

In my earlier 'voyage of discovery' over the winter of 1981-1982, I discovered small operators. These were the residual 143 small sawmills and other products catalogued in the Saskatchewan forest industries directory, 1979 (DREE). On a November day, I held a tailgate meeting, in the forest north of White Fox, SK with small sawmill owners. Walter Bailey was with me. This contingent of small operators was angry, and some almost hysterical in their frustration. A significant patch of prime white spruce blowdown, located in Saskatchewan Forest Products Corporation interest area, was denied to them. These were the relics from the White Spruce Lumbermen's Association, left over from the days of the Saskatchewan Timber Board. What made matters worse was that Saskatchewan Forest Products Corporation had no intention of salvaging that blowdown. A

162

further irritant reduced the chance of a wood supply for small operators in the Hudson Bay district. The Hudson Bay forester and his operations unit were in bed with Simpson Timber and Saskply.

One glaring example of unequal treatment involved John Farber, a small sawmill operator in the Porcupine Plain Forest, Hudson Bay district. He wished to salvage some prime white spruce surrounding a beaver flood. Hudson Bay district refused to consider the application. John Farber contacted me in Regina. I organized a meeting on site with John and the Hudson Bay resource officers. It was a cold day in February, a brilliantly blue sky, and trees coated in hoar frost. All concerned walked the area, the resource officers pitching the case the trees were still alive and they should be left standing. I went up to a huge white spruce on the edge of the beaver flood, looked up to see the branches covered in hoar frost, and kicked the tree. Instantly, the dead bark from the top to the bottom fell in a heap at the base of the tree. That tree, and many others, was dead. The beavers had drowned them. I asked the officers to issue a salvage permit, not only for the dead trees, but also for the residual green ones that would die in the next year or so. I earned lifelong antagonism from the Hudson Bay staff from there on in. Especially, when the operations forester in Hudson Bay transferred to the forest operations manager's position in Prince Albert sometime later. Vic Begrand had left to run the Canadian Forest Service district office in Prince Albert.

The small sawmill operators produced lumber from the biggest trees using a P.A. No. 2 portable mill designed by Prince Albert Foundries. This consisted of a 50-inch sawblade equipped with 3/8-inch saw teeth and power takeoff. Most farmed in the spring and summer, then sawed logs in the winter, when they could get any. I expand on this later. See 1985 to follow.

The following summer I asked forest operations in Prince Albert to issue a timber permit for a stand of large jackpine timber. A small operator needed it to fill a rail tie order. He had a long history of filling orders from CN rail in Edmonton (R & T Trucking based in Prince Albert). This timber was within the PAPCO area of interest. The ties were shipped, but this was his last order. From then on, the pulpmill made sure that jack pine of rail tie quality was logged and pulped.

Wood treatment plants (preservers)

Since 1886 or thereabouts, the forests north of Prince Albert produced the rail tie needs for railways across the Prairie. The demand was significant. Small operators and Indian bands sawed up the pine, then took it to the Domtar treatment plant on 15th St E in Prince Albert, which was built sometime later to save shipping untreated ties to Edmonton. The 1972 Northern Forest Research Centre economic analysis showed the presence of six wood treatment plants in Saskatchewan. By 1981, the Domtar Chemicals treatment plant in Prince Albert was extinct.

"Sask Forest Products (SFP)/Domtar: A wood treatment facility which operated within the city of Prince Albert from the 1930s to 1976 resulted in heavy contamination with creosote and other wood preserving chemicals, including pentachlorophenol. The problem was compounded by the installation of a storm sewer across the property in 1979, which acted as a conduit for contaminants to migrate into the North Saskatchewan River. The Centenary Fund supported a cleanup begun in the early winter of 2002, continuing into early 2003. The fund contributed $1 million toward the remediation of an area adjacent to the SIAST 12th Avenue Woodland Campus access road and an old surface drainage channel. In 1994, a partnership involving the province (through SFP's successor, Crown Investments Corporation), CN Railway and the City of Prince Albert spent another $3 million to clean up the SFP site and reinstall the sewer so as to arrest the flow of contamination to the North Saskatchewan River. Saskatchewan Environment is currently meeting with the city of Prince Albert to determine what additional monitoring and/or subsequent remediation will be carried out in the future." (Source untraced).

It was a remarkable use of Centenary funds. Most provinces built community infrastructure with their allotments.

However, the Sask Forest Products treatment plant on the north side of the North Saskatchewan River, within the city of Prince Albert, stayed open until the late 80's to early 90's.

Just before Christmas 1983, I called on Vic Begrand, the forest operations unit manager in Prince Albert (he took over from Frank

164

Flavelle) to come to Regina and produce a wood supply protocol for the fencepost wood preservers in Saskatchewan. This was an industry eternally ignored by government economists, but it supplied farmers throughout western Canada, and down into the States with barn poles, corral fencing, fenceposts and poles, all made with treated jack pine, as well as second-cut slabs for windbreaks. PAPCO throttled the growth of this industry. It did all it could to block access to the fencepost timber. However, Vic redirected some of the old Domtar limits into the right hands. Since post cutters received a piece rate for each post, this produced a labour-intensive process. Aallcann Wood Suppliers, for instance, had 100 post cutters on the payroll, and all from the north. Although, Aallcann did not benefit from this redirection. Later, I estimated that fencepost demand was in the neighbourhood of 7 million pieces annually, but timber supplies limited this to slightly over 2 million. Oh, the obscene power of Treasury Board and that board running PAPCO.

Also by Christmas 1983, I submitted our proposed action plan for forest management in Saskatchewan to the government's 'Executive Council'. The thrust of this document was to balance the needs of the big industries with those of the small operators. The big guys produce larger taxes, but the little guys produce more jobs on a-ton-by-ton basis. The big guys concentrated wealth, and the small guys dispersed it amongst sixteen communities along the forest fringe.

1984

Wheat and other cereals are harvested on an annual basis from the same hectare. Commercial timber crops supply a multiple product array after logging: peelers for the plywood plant, sawlogs, and pulp – all from one hectare. But the hectare rotates over a 100-year cycle (rotation). The pulpwood component comprises 40 percent of the volume removed from each logged area. PAPCO played some dreadful games with both small operators and the two Crown sawmills by refusing to take delivery of byproduct pulp timber. The absence of an integrated forest industry played havoc on Saskatchewan's forest economy. Raw wood costs, both to the pulpmill and to the rest, were inflated because of the timber wasted but paid for. PAPCO hated the Crown sawmills and did all it could to

165

make life in the Saskatchewan forests difficult. If it could, it logged a sawmill stand before going after a smallwood pulp stand nearby.

Thus, my action plan for the sustained management of an integrated Saskatchewan forest economy was an answer to Allen Blakeney's forestry quandary. For one, Dale Botting in Executive Council understood the desired outcomes proposed. The Devine government considered it to be an NDP plot and fired me as of February 28. Although, the true villains were housed within Department of Finance and Treasury Board. None of this was political in reality. It was just 'turf'. I was left with no income, three teenagers and my wife, not to mention the mortgage.

I dropped in to visit with Honourable George Mcleod and told him that I was not leaving Saskatchewan, and that I could do more for forestry outside government than inside. It was a cordial separation. I finally received a $10,000 separation. I liked George.

Ω

Chapter 15:

Silviba Services Ltd start up

I started up Silviba Services Ltd on May 11, 1984, a consulting forestry company registered in Saskatchewan. I loaned the $10,000 to a neighbour in worse trouble than myself, and never saw it again.

My first contract was to explore a wood supply for a proposed charcoal briquette factory in Moose Jaw.

Later, in the spring, I received a government contract to mediate the fight between PAPCO and the two Crown sawmills over road user fees. After deliberating with all parties, I presented my decision. It also included a protocol to remove any further road user grievances by either party. PAPCO had to pay a settlement price to the Crown sawmills. These in turn understood the need for new rules and prepared themselves to act accordingly. When I presented my report to the woodlands manager in his office at the pulpmill, I received a serious slap on the face by this irate individual. There were witnesses, also. All water off a duck's back, though.

Federal Department of Labour grant
In June I obtained a federal Department of Labour grant to visit Métis communities in northwest Saskatchewan, as well as Meadow Lake Tribal Council (MLTC) and Prince Albert District Chiefs (PADC). My mission was to spread the word about opportunities within the Saskatchewan forest. The options were in developing small community forestry businesses, or through logging contracts. It was a great trip. I slept in kids' beds while they were relegated to the chesterfield. I attended community meetings in Ile a la Crosse,

Buffalo Narrows and Beauval. On my way home from Meadow Lake I attended a meeting arranged by Meadow Lake Tribal Council, a consortium of seven Indian bands. I met chiefs and key white staff. What I failed to realise was that these folk were in the gun sights of Fortunato Pacios-Rivera, an independent rogue forester. He was in the room as I explained the business opportunities streaming from the proposed PAPCO FMA. My visit confused them. They were torn between two messengers. It took a few years, but eventually MLTC countersued Fortunato for fraudulent invoicing. Pacios-Rivera's bargaining tool was a hunger strike on Parliament Hill in Ottawa. Much later, he settled the matter out of court. One of the conditions was that Fortunato was to remove himself from Saskatchewan forthwith.

The Prince Albert District Chiefs (PADC) office across from the new city hall, on 11th Street E. was my final contact. This represented a consortium of twelve bands north and east of Prince Albert. It was an August afternoon. I met with Phil Parr, director of Economic Development, and Chief Milt Burns. I used a flip chart to outline a sketch map of existing PAPCO areas of interest. PADC and the Indian reserves should join Forest Management Agreement negotiations. They could then create economic benefits from forest resource management within their traditional land area: jobs, business startups and skill development. Phil was a former social worker. My brief to PADC was so effective, that Phil figured on doing all the follow-up himself. Thus, this strategic project died before it ever got off the ground.

Lakehead University, Thunder Bay, ON
In August, I drove the old suburban inherited from IFS to Thunder Bay to teach both forest fire management and forest management at Lakehead University. I took the portable forest fire simulator with me. Once there, I worked incredibly long hours, and rewrote the lesson plans for my three courses. I suffered significant sabotage from a teaching assistant who thought he should have had the job, and I lost eight pounds. The downturn in the forest economy Canada-wide affected the students' 'keenness quotient'. They were only present to tap into government labour grants. In addition, the timber growing on the Precambrian shield behind Thunder Bay would

depress anyone. Oh yes, my office was directly over the boiler, thus the temperature in the office was never less than 85 deg. F.

Two fortunate events presented themselves after Christmas. First, I had permission to attend a week's course in Sault Ste. Marie involving the unveiling of the new 'Canadian Forest Fire Behaviour Prediction System'. I had access to the foremost fire management experts throughout Canada, including Brian Stocks, the architect of Canadian fire behaviour prediction. From then on, I could articulate forest behaviour rather than feel tongue-tied and inadequate. If only I had had this skill while on the witness stand during the Muchalat fire trial in 1968. So, I wrote this material into my forest fire management lectures after Christmas.

The second involved fire simulator training for both my students and the regional Ontario Forest Service fire folk. The forestry Dean had managed to get dedicated space in a classroom set aside for the setup of the simulator. It was an excellent training site. Ontario Forest Service folk sent trainees and they took part fully in the exercises. It was 'win win'. At the end of my teaching contract in May, I gave the simulator to the forestry school.

One of God's favorite exercises on the simulator, was to start a lightning fire within 300 feet of the shoreline on an island. An initial attack crew in a float-equipped airplane with room for five crewmen, pumps and hose, chainsaw and handtools would offload on an adjacent beach. The fire, burning in a mixed black spruce-jackpine stand would start to creep north due to a south wind. Instead of attacking the fire head and removing the fuel, the crew always opted to set up the pump first. By the time water was coming out of the hose, the fire had travelled beyond the reach of the hose nozzle. This was a classic mistake.

Before leaving Lakehead University, the student grade for my teaching performance came in as a 'fail'. I put this down to students' resentment of their workload, and sabotage by the disgruntled teaching assistant. Back to Regina.

Ω

169

Chapter 16:

Silviba Services Ltd at-large

In May 1985, I set up a bank account for Silviba at the new Bank of British Columbia branch in the Gateway mall in Prince Albert. I borrowed money to lease a Ford Ranger and set up a line of credit. This did not last long. The Bank of British Columbia closed and I had to find a replacement bank. The Toronto Dominion Bank on Central Ave took over my account. A gracious act.

I moved my workroom from my apartment to downtown, into the basement of the Prince Albert District Chiefs. The same two sawhorses and the sheet of plywood. This time I had a Company phone. I used First Page Business Services as my answering and secretarial services. Wendy Wiebe (Wiebechick) and two other church women had set up this business on the ground floor of what is now the Gabriel Dumont Institute, on 12th St. West. Several other small businesses used this service also, and some had offices within its premises. Altogether, it was an ideal arrangement for Silviba and the others. Wendy's office was only up the alley and around the corner from Silviba's workroom. Also, in 1985, I listened carefully to determine who had me fired from government. The Member of the Legislative Assembly for Christopher Lake, Paul Maher, a plumber, sat on the special Crown board that managed PAPCO. He bragged that he was the instigator. *C'est la vie.*

Small sawmill operators

I discovered small operators once more. Between fencepost treatment plants and small sawmills, the total timber volume under permit approached 8 million board feet a year. But they were lucky to receive half that. Yet, the accumulated orders amounted to 60 million board feet of lumber. On the other hand, the Crown sawmill would struggle to achieve 20 million board feet of production; although, mill capacity approached 60 million board feet. Crown corporations had a monopoly on the sawlog supply. Forestry Branch's role in the timber allocations was limited to one-year permits. Thus, financing of operations was also a limiting factor. Banks required at least a five-year permitted timber supply – preferably ten. Operators needed advances for access road building, maintenance and repairs, log inventory, as well as a regular line of credit.

Also, small operators logged pulpwood as a byproduct. Forestry Branch considered this residual pulpwood as a commercial product. Yet, the pulpmill refused to purchase any. Fencepost operators created the most byproduct pulpwood because of the nature of their business. The manufacture of black spruce rails also created more pulpwood.

I encouraged small operators to form the Saskatchewan Council of Independent Forest Industries (SCIFI). In the short term, this solved nothing. Small operators were intensely competitive, and cooperation was scarce. Honourable Colin Maxwell, minister responsible for the Department of Parks and Renewable Resources, attended a SCIFI meeting at the Marlboro Hotel, downtown Prince Albert. He promised it a volume of 15,000 cubic metres (300 truckloads) annually of byproduct pulpwood. Shortly after, he was run out of government and quit. Silviba did the fieldwork and follow up paperwork for several small operators in the annual timber permit applications. I will recommend a fairer approach to forest policy and small operators in the final chapters. After all, even the Weyerhaeuser family started small, on their way to becoming a major forestry corporation.

Small operators were a pain-in-the-neck for Forestry Branch. The pressure for timber was unrelenting. The administration required bush inspections, and these were the responsibility of another branch and its director. It did not help that Honourable George McLeod planned to change the name of the resource officers to conservation officers, and then provide uniforms and side arms. There is no feeling more intimidating than to have a conservation officer equipped with a sidearm stalking across your slash, accusing you of pirating timber.

His GPS told him that your logging was in the wrong place, when, in fact, he turned out to be in the wrong block of timber. Strapping a sidearm to anyone changes behaviour, creates aggression and everyone else is judged guilty. At least Bill Klassen and I agreed over the conservation officer issue. The only saving grace in Prince Albert was the six conservation officers wedged into one tiny office space only had one truck to share between them. On top of this, small operator lumber output was at minimum levels. One conservation officer could have kept in touch with every operation. Regardless, I deplored the circumstances in which they did their jobs.

Fencepost operators

Fencepost treatment operations were equally starved of raw material. The pulpmill would change its logging plan overnight to grab a stand of fencepost quality jackpine close to roadside, rather than see it go to a treatment plant. There were three treatment plants in Prince Albert. I had ensured that two of these had longer term wood supplies in my director days. However, the third operator, Aallcann Wood Suppliers, was left out, and had to scrounge for every stick of wood. It needed 30,000 m3 of post material annually, exclusive of a pulpwood byproduct quota. This, it needed to direct to the pulpmill.

A subset of the fencepost business was the 'peddlers'. These small truckers circulated amongst their farm clients all over the province delivering fenceposts directly from the treatment plants. They also delivered lumber, rails and second cut slabs for corral fencing. At times, a shortage of lumber products meant they had to go south with a light load.

Aallcann attracted conservation officers like bees to pollen. The major irritant was byproduct pulp logs. The fencepost cutters were paid by the post. Poor quality material was not paid for. This meant that byproduct pulpwood volume accumulated as the cutting progressed. The conservation officers attempted to fine Aallcann for leaving merchantable timber behind after logging. To be merchantable, a log would need a market. Since the pulpmill refused to accept it, the material stayed on the ground. Again.

173

Later, Aallcann hired its own forester. His first job was to set up a timber scaling routine as a basis for the payment of Crown dues. The government could have given the fencepost timber away at a much saving in administrative cost. Or, collected a value-added tax based on sales receipts. However, government officials involved made sure that an elaborate sampling system was in place to create work for themselves. Since Aallcann shipped in loads of posts 24/7, conservation officers set up a webcam in a tree one time to catch loads bypassing the scaling yard on their way to be offloaded. Unfortunately, all that overtime resulted in no data collected – no battery in the camera. All this, in spite of the fact that the Aallcann scaler was government licensed. The whole situation smacked of 'turf', given that scalers were the responsibility of Forestry Branch, not the district staff.

The one thing that relieved Aallcann's wood supply crises was burnt timber originating from large forest fires. It was free of Crown dues for the most part. There was lots of it as a rule, given the frequency of lightning fires. Because the material was dry, Aallcann treated in winter as well as summer since the treat entered the wood cells easier, and the bark peeling process was faster. The bark fell off as each post fed the peeler. Also, burnt timber was not usable as pulpwood.

Today, Aallcann's treatment plant in Prince Albert survives on a wood volume of 60 tonnes/ha. This is enough to run 2 fencepost peelers for 2 days back at the treatment plant. In 1986, there were five plants of this nature in Saskatchewan, plus the Crown facility that replaced Domtar in the 70s. The Crown treatment plant was decommissioned in the late 80s, both leaving behind sites contaminated with creosote. This treatment material was outlawed in the 80s and replaced with CCA (chromated copper arsenate).

It is also noted that Lehner Wood Preservers with its plant at Redwing and Vermette's with its plant at Spruce Home rounded out the treatment plant industry in Prince Albert. The 'dippers' have faded out of the picture due to the banning of creosote as a preservative.

Sawmill economics vs farming

The people of Saskatchewan, through their provincial government own the forests of Saskatchewan. The Indian people were kicked off their traditional lands in 1930 through the Natural Resources Transfer Act. Whereas a farmer gets a wheat yield of 36.7 bushels/tonne, worth $277.67, an industrial logger gets a yield of 170 m3/ha, or 1 cubic metre/tonne, worth $52.50/tonne delivered to the sawmill yard from a timber stand of sawlog quality. In 2014 the delivered wood cost to the sawmill was $52.50/tonne, Crown timber dues extra.

Next, the farmer's value at 62.5 bushels/ha = 1.7 tonnes/ha = $472/ha. The logger, on the other hand, generates 170 times $52.50, or $8,925/ha. Then add the Crown royalty portion of $510/ha[35].

Last, on a 'jobs' basis, a farmer combining 160 acres (65 ha) creates employment for 3 people on a 12-hour shift, or 4.5 people on an 8-hour shift basis. Compare this with a small operator, who logs and hauls 12,500 fbm/day using 7 men. But, this goes on every working day for 8 months per year, whereas the farmer may head for Arizona once his harvest is completed.

And that small sawmill goes on to process those delivered logs here in Saskatchewan, creating a further 7 to 32 jobs and a total cashflow of $15,742/ha, not including indirect employment – factored at 1.4 times by some economic analysts.

I am not familiar with the cattle industry. Someone else can make the comparison based on tonnes vs hectares. Then there is the mixed cattle/ cereals producer.

Bottom line - the forest industry adds significant value to the Saskatchewan economy. This used to be spread throughout sixteen communities across the Parklands. Not today, though.

Small operators add 25% to the provincial allowable cut. This is the sustainable yield of sawlogs, pulp, poplar, poles and fenceposts available over the lifetime of a forest. Why? Large operators such as Weyerhaeuser cannot afford to log forests under 20 ha in area. Because of mechanical harvesting with all of its attendant expenses and overhead, Weyerhaeuser bypasses this 'chicken feed'. It leaves behind many small bluffs of commercial timber in its day-to-day operations.

The small forest operator had no expensive overhead. He could move in his skidder, bunkhouse and D7 tractor for building roads for some $8,000, round trip. He mopped up small patches of sawlog material, including windthrow. His main source was the salvage of forest fires, of which there were several over the years.

Saskatchewan small operators in the mid-80s had full order books. Larger sawmills in western Canada were down to one shift – or shut down – during the worst lumber market crash since the 30s. Why? Saskatchewan small sawmills sawed specialty products to order. The dimension lumber produced was full size, not planed down, and was air dried. Items such as bridge timbers, posts used to support highway signs, bridge decking, full 2X4, 2X6, 2X8 and 2X10 material in all lengths up to 20 feet. Add in the manufacture of 4X4s, as well as second cut slabs for corrals, and poplar 4X4 material used to place under pipelines during construction. Each sawmill had its own connections through lumber brokers for its various products. Much material went east to remanufacturing plants in Manitoba, Ontario and Quebec. Saskatchewan white spruce produced 98 percent no. 2 and better in lumber grade. It has strength characteristics comparable to Douglas fir. Weyerhaeuser's building supply depot in Winnipeg also bought this material for resale.

Small operators created cashflow and jobs throughout the Parklands between Pierceland in the west and Hudson Bay in the east. Large, integrated forestry corporations, on the other hand, were concentrated in Meadow Lake, Prince Albert or Hudson Bay. The 1979 Directory of Wood Industries in Saskatchewan listed some 249 forest industries[36]. Most of these existed during the post war Saskatchewan Timber Board era. Since the birth of forestry Crown corporations in the 60s, and latterly, Weyerhaeuser Canada, together with the inherent animosity for small forest operators by the Saskatchewan Forest Service and its predecessors, today's small sawmills can be counted on the fingers of one hand. Economic genocide. It is a cashflow component sadly missed by local fuel suppliers and RM administrations, also by farmers who operated sawmills in the winter.

In fairness to government forest agencies, the legislation and regulations did not allow for the fair distribution of the allowable cut among all eligible forest industries. I blame Treasury Board and Executive Council in Regina. Has anyone heard of diversification?

1986

Honourable Colin Maxwell gave me two assignments: The first was in response to a visit I made to the minister on my return from teaching at Lakehead University. I briefed him on the new forest fire behaviour prediction system, wrote a report. This generated the second assignment.

In cooperation with the director of the province's forest fire management branch, I put on a week's course for all field directors in Prince Albert. They learned the essentials behind the prediction of forest fire behaviour. The Canadian Forest Service fire research expert from Edmonton provided a briefing on the Canadian Forest Fire Behaviour Prediction System. The teams fought a fire with given inputs using actual maps and conditions. It was a good week. It did not take long to see the results as the provincial forest fire management system upgraded procedures.

Weyerhaeuser Canada, Feb 1986

In February, I received a phone call from Weyerhaeuser's chief forester in Kamloops. It was a Friday. The conversation lasted for hours. At its conclusion, the chief forester requested a further phone meeting on the following Monday, when the vice president could join in. After this Monday conversation, that lasted the better part of four hours, we all agreed to meet together in Prince Albert the next week.

They showed up in my sawhorse workroom, not really reflective of a prosperous consulting forester's lair. We went upstairs together and had lunch in the nearest restaurant. Bill Gaynor, executive vice president, and Steve Tolnai, Weyerhaeuser's chief forester, sat across from me in a booth. Steve had worked with me back in the Nimpkish days. They started the 'good guy-bad guy' routine. Bill supported my message that small operators needed a

177

secure wood supply, whereas Steve intimated that Weyerhaeuser shareholders would not care one whit whether these folk lived or died.

Later that spring, Honourable George McLeod flew into Prince Albert specifically to visit with me. We met at MLA Lloyd Muller's office. George informed me that the government had merged the pulpmill with the Crown sawmills. This was one key component of the forest policy updates presented to government in my director days, but that only got me fired. In September, Weyerhaeuser and the Saskatchewan government signed an agreement that transferred ownership and timber rights. This included signing the Weyerhaeuser Forest Management Agreement in September 1986. No more forestry Crowns. Finally, I had my bittersweet victory. However, small operators came out of this with a token volume of pulpwood and severely limited timber allocation.

Gord Frey in Forestry Branch was the forester assigned to parceling out small operator timber in the Weyerhaeuser era. It was a gruesome assignment. The matter of byproduct pulpwood had not gone away. The pulpmill woodlands group was kept on by Weyerhaeuser, so attitudes remained unchanged. On top of this, small operators seeking white spruce sawlogs were assigned to mixedwood timber stands. Weyerhaeuser was now harvesting a high percentage of poplar for paper-making. For small operators this created a new byproduct, poplar logs. And Weyerhaeuser made Gord Frey's life even more difficult by refusing to take byproduct poplar more than the bare minimum. And, this was the Weyerhaeuser Canada forest management template. Those oversize logs that the pulpmill had buried from time to time, which then caught fire from spontaneous combustion, were no longer buried. They were sold to small operators after Weyerhaeuser bought the mills.

Weyerhaeuser inherited the pulpmill management. Forestry Branch was in a continuous state of stress as this management leaned on government foresters to minimize permitted pulpwood volumes, both softwood and poplar, within sawlog or fencepost timber permits.

One particularly devastating incident comes to mind. R&T Trucking held a timber permit on the Montreal River close to the Besnard road, west of La Ronge. The local government forestry officer accused R&T of pirating timber – logging both sawtimber and poplar volume in excess. The officer emptied the contingent of clerical staff into the bush to measure every cut stump. They fined Harry Romanchuk $30,000 and shut him down for a year. Silviba went in there later and also recorded stump diameter and species logged. Harry had logged poplar that he shipped to Weyerhaeuser, and white spruce that he transported to his own sawmill. Silviba's crew, myself and another forester on Silviba's payroll, remeasured all stumps, but we separated the white spruce stumps from the poplar stumps. The government had marked all stumps the same colour. Silviba then calculated the net log volume from a sample of trees standing adjacent to the logged area in dispute. The white spruce volume recorded by Harry's load slips matched the Silviba estimate for white spruce sawlogs and pulpwood closely. Yet, the government still accused R&T of pirating timber. Weyerhaeuser had incited the attack then hid and watched the outcome. He had no money to fight the government action, so shut down for a year. The government Forest Operations office in La Ronge denied that it had committed errors in the audit. An example of one of Weyerhaeuser's pressure tactics.

At the request of Weyerhaeuser woodlands management, on another of R&T's winter cuts, Silviba compared the volume of timber logged under permit to the volume of lumber legally sawn and recorded in R&T's mill book. I compared the difference between the permitted volume logged via the load slips and volume sawn. The difference between the total volumes on the load slips compared to the mill book record was within 1% of the timber volume allowed.

A Prince Albert resource officer harassed Provincial Forest Products Ltd, a small operator with a sawmill in the rural municipality of Buckland, for hiding small sawlogs in the pulpwood deck at his bush landing. The government took him to court at Smeaton. The operator won the court case since the suspected logs were not sawlog quality (crooks, sweep, and decay). The officer was posted to Uranium City for a time after that as a penance.

I enjoyed my forestry service work with small operators. Silviba did cutplan submissions to the government, laid out logging roads, ribboned out the streamside leave strips and mapped out the logged areas. Silviba also liaised with Forestry Branch. I caught Weyerhaeuser one late October burning decks of good poplar in R&T's old cut area. Harry Romanchuk had originally set aside this wood for transport to the pulpmill. He had paid for the logging and decking, yet Weyerhaeuser refused to accept it. Dead silence from the conservation officers – and from Weyerhaeuser. Government officials showed up in Weyerhaeuser's logging operations only when they had lost their way. Self-regulation anyone?

Saskply job

In this earlier era, the Crown plywood plant in Hudson Bay hired Silviba to do a study on waste created by the log peeling process. In the winter, a steam chamber thawed the logs. From there, a lathe peeled them into long sheets of veneer. As these left the lathe, they flowed over a table. An inspector cut out the defects with a guillotine. The study team measured all waste veneer dropped into the conveyor leading to the steam generator on a two-shift basis for one entire week. The balance then fed into the dry kilns. This reduced the moisture content to 12% prior to lay up. The dried veneer was covered with a glue as it passed into the press. It came out as plywood.

Because of this study, Saskply bought two new machines from Germany. These used the waste veneer, thus increasing the production of plywood by 28 per cent with no increase in fixed costs. Not only that, but Saskply management used poplar veneer as a core stock with spruce veneer for the external surface. – much cheaper to produce. Perhaps this also solved Allan Blakeney's quandary about how to make Saskply more effective and efficient.

Manitoba Oriented Strandboard (OSB) plant timber supply

In July, a consortium headed by Gary Bowell, former president of Weldwood Canada, hired Silviba. The terms of reference included white poplar (trembling aspen) around Swan River, Manitoba. They had money to build an oriented strandboard (OSB) facility. Where should they locate it? How much timber supply was accessible? Silviba completed a survey of all poplar stands between Lake Winnipegosis and the Saskatchewan-Manitoba border east of Hudson Bay, SK. Manitoba Natural Resources in Winnipeg and Dauphin aided with maps and printouts of the timber supply. Silviba developed a wood supply plan capable of feeding the new plant with the needed volume and quality of logs at the right price.

But, to realise the full potential of the proposed plant, an extra volume of poplar was needed from eastern Saskatchewan as well. I conferred with both Saskatchewan Parks and Renewable Resources and MacMillan Bloedel, who ran an OSB mill in nearby Hudson Bay. Since M&B used the same wood profile as the proposed mill in Swan River, it was essential to receive their approval before continuing. I then added a Saskatchewan ingredient to wood volume needed. This improved the viability of the proposed plant in or around Swan River.

I had two last jobs on this project. The first was to produce a delivered wood price estimate to the proposed plant. To do this, I had to settle the location of the plant.

Mafeking, just down the road from Swan River, provided the shortest average log haul distance from the forest to the proposed plant. Fortuitously, the proposed site was close to rail and natural gas. And this is where it was built.

Looking back

On Feb 17, I declared that I was now off unemployment insurance. Before this, I had worked for Canadian Executive Service Organisation (CESO), an agency utilising retired executives to provide expertise to embryo projects of all kinds. They reimbursed me for expenses incurred during my Indian involvement. As the jobs started to create cashflow, Silviba hired two technicians, Keith Macauley and Gerald Brahniuk. Brian Firby, a forester from Saskatchewan Forest Products Corporation, joined the firm in June. Keith's dad, in the past, was a Member of the Legislative Assembly representing the north. Also, Silviba had developed a crew of forestry field assistants from both Peter Ballantyne and Montreal Lake reserves. By year-end, the company operated three vehicles: two leased pickups and a suburban.

Silviba had completed or arranged for 16 projects in the forestry domain. I spent time with negotiations and business planning triggered by various funding agencies and economic development directors, especially in the Indian world. Northern economic development in Saskatchewan meant sharing the risk. To get all the needed startup funding, a band would need to access up to three or four sources from various agencies – mostly federal. The time consumed to stitch together a financing plan was horrendous. These were redone up to five times before one of the agencies backed out, leaving behind a mountain of client frustration. Rule changes amid negotiations, an agent hiding behind technicalities, all produced much bureaucratic fuzz, but few results. However, a few government economic development agents did contribute their expertise significantly. After I had redone the Montreal Lake five-year forestry business plan five times, I realised that Bob Cannon, of the involved agency, had taught me how to do business planning. Unfortunately, a few agents were lemons. In one case, Silviba did a five-module training package for Canadian Employment and Immigration Commission in Prince Albert. As soon as this agency approved and accepted the training plan, the government contact responsible for the project quit and went out on his own as a 'training consultant', and took the Silviba package with him.

Chapter 17:

Silviba Services Ltd and Montreal Lake Indians

Early on in Silviba's association with Woodland Cree people, I asked Chief Roy Bird of Montreal Lake Indian Band (MLIB) for advice on how to reference his indigenous band members. His reply: "Call us Indians". The community of Montreal Lake was exactly one hour north of Prince Albert on Highway 2 north. This was close to a mainline log truck haul road connecting to the pulpmill (PAPCO). In fact, the asphalt top coating on the inbound lane to Prince Albert was twice the thickness of the lane taking empty trucks on the back haul to the logging operations. Montreal Lake was at the heart of the proposed Forest Management Agreement area, and the pulpmill wood supply. The band had earned a satisfactory reputation for tree planting on the PAPCO lease. The pulpmill woodlands management had not considered any deeper relationship. This lease also happened to contain a high proportion of the MLIB traditional landbase.[37]

When I discovered Roy Bird in his office at the old band office on some spring day in 1986, Roy quickly launched into his vision of a massive particleboard factory. I did not connect that message with Fortunato Pacios-Rivera right away. However, sometime later, I came across Fortunato's stack of timber inventory reports for Montreal Lake reserve completed and paid for by Canada Employment and Indian Affairs. Little did these folk know that Fortunato used Scandinavian forestry students as summer field help? Also, he used Swedish volume tables for the calculation of timber volumes on reserve. None of this registered with Roy, who received one of

Fortunato's plaques for 'excellent forest management'. Also, that stack of reports lacked the associated forestry maps produced and paid for, but the band was unaware of these. Neither was the funding agency. Where there was knowledge, there was also power, and Fortunato kept the maps to himself. Without these, the associated numbers meant nothing.

Chief Roy arranged for Ernest Halkett to show me various locations for forestry potential on the reserve on his Elan skidoo. Ernest let me drive across a hummocky muskeg, but I ejected him into a snow bank en route. I remained as a passenger for the rest of the way.

Chief Roy went on to become Director General for Indian Affairs in Manitoba and Saskatchewan. He was based in Winnipeg.

The new Federal-Provincial Forest Management Agreement contained program funds for timber inventories on Indian reserves. This expanded into tree thinning and reforestation programming. The Canadian Forest Service office in Prince Albert managed this in Saskatchewan. Although INAC (Indian and Northern Affairs Canada) controlled the budget.

With forestry funding available, MLIB redid the forest inventory mapping on reserve with guidance from Silviba, and completed the five-year forestry business plan developed by Silviba and Frank Hart's company in Regina.

Ω

Extract from the Montreal Lake Cree Nation – with permission *from Upress u regina.ca.*

"Montreal Lake Cree Nation

Following their move from Grand Rapids at the northwestern end of Lake Winnipeg to the Montreal Lake region in the mid-1800s, the band signed an adhesion to Treaty 6 on February 11, 1889, under Chief William Charles, and its reserve was surveyed at the southern end of Montreal Lake in 1890. The Little Red River Reserve, surveyed in 1897 for the joint agricultural use of the Montreal Lake and Lac La

184

Ronge Bands, was divided between the two in 1948. Little Red River Reserve (106B) is situated 39 km north of Prince Albert and currently hosts a sub-office with a band hall, health centre, day care, group home, and confectionery/gas outlet. The first Hudson's Bay Company post, constructed at Montreal Lake in 1891 in response to the establishment of the reserve, became the freighting depot for goods coming in and going out of the region. Until the late 1920s band members found ready employment as labourers and suppliers. Horses gave way to caterpillars, and airplanes and trucks started arriving by the end of the 1920s. An all-weather road was completed in 1937; whereas it significantly reduced labour employment, it also created a niche in early tourism as summer tourists began flowing into Montreal Lake in significant numbers. Forestry, recreation, and tourism remain important to the economy of the band. The Montreal Lake Development Corporation (1985) has four subsidiary band-owned companies, and shares with the Lac La Ronge and Peter Ballantyne bands in three others. The Montreal Lake, Little Red River, Timber Bay, and Weyakwin communities form the Montreal Lake Reserve (8,288.8 ha), on which 1,773 of the 3,108 band members live."

Christian Thompson

©2007 University of Regina and Canadian Plains Research Center Terms of Use located at: http://esask.uregina.ca/terms_of_use.html

The Little Red River Reserve, surveyed for the joint agricultural use of the Montreal Lake and Lac La Ronge Bands, followed in 1897. This reserve also contained superior timber stands of white spruce. These form the basis of subsequent claims for compensation from timber stolen by Prince Albert Lumber Co. between 1905 and 1919.

In its own way, the band moved ahead. Montreal Lake Indian Band became Montreal Lake Cree Nation (MLCN). This registered a new company, Montreal Lake Development Corporation (MLDC). This Corporation built a new mechanical and carpentry shop and offices beside the equipment storage yard. The band built a new band office with meeting hall. As well, a new school and a hockey rink added to the quality of life on reserve. Apartments for singles also came to be over the years.

The chopsticks project

Silviba's first formal connection with Montreal Lake Development Corporation (MLDC) occurred in the winter of 1986-7. A company from offshore started to seek a viable poplar raw wood supply suitable for the manufacture of chopsticks. Plainly, the demand for these was growing exponentially as the preference for disposable chopsticks caught on. Montreal Lake Development Corporation bought several chainsaws and refurbished an old school bus. We bought camping supplies for seven young men, as well as safety gear. We needed to collect data on the quality and quantity to develop a raw wood supply inventory. This, of course, would serve as a key part of the business plan to follow.

The preferred tree species was trembling aspen (*Populus tremuloides)* found on the reserve, as well as at other locations, over a broad area. Candidate trees were tall, straight and of superior girth. I developed a sampling plan to collect key measurements:

- Diameter and total age at stump height
- Diameter of tree each two metre interval up the tree to the base of the tree canopy
- A cookie from the top of each of the 2-m sections in each tree sampled was sawn off and brought back to MLDC.
- Total age at the stump and diameter of the top of the last section
- Bark thickness was measured as well, since tree diameters are always 'over bark'.

The boardroom at MLDC served as a workroom. The crew sanded down each cookie. They counted the annual rings and calculated a volume deduction factor for knots. Chopstick material had to be free of blemishes. Their data was immaculate. I slept under the boardroom table for a week while supervising the work.

We then moved the sampling off-reserve. The crew travelled in its bus and towing Senator Allen Bird's double track skidoo, complete with its eight-man sled. We covered an area as far east as Greenwater Provincial Park, and as far north as the very south end of Lac La Ronge. I relate two of the several incidents from that February:

186

At one location, we camped just off Highway 6, in an old gravel pit. It was cold. Easy Rider drove Allen's skidoo along the top of this esker – a sharp drop on either side, not too far north of the Saskatchewan River which was iced over. I sat behind Easy Rider, with the rest of the crew packed into the sled. The steering pin broke. The machine went over the edge, trailing the sled that stayed upright. I rolled off, as did Easy Rider. Everyone helped get the machine back up on to the trail, repair the steering and rescue the broken windshield before continuing on our way. That is how 'Easy Rider' got his name. His last name was 'Bird'.

Camping there above the river, we could not get warm. We were north of Melfort in the Fort a la Corne forest – just north of the bridge crossing the North Saskatchewan River on Highway 6. It was February. The boys just had blankets and were very uncomfortable. They put me beside the stove. My job was to keep the wood heater going. None of us got much sleep that night. The next morning we had to light a small fire under the oil pan to warm the crankcase oil in the bus before we could get it started.

We then drove up to Wapawekka Lake, just south of Lac La Ronge. We used a trail that linked Wapawekka Lake with the extreme south end of Lac La Ronge. We skidooed onto the ice and found ourselves awash in freezing water. Our objective was a bluff of good poplar on a point at the south end of Lac La Ronge. We built a brush mat on the ice to get the skidoo turned around and facing the shore. Everyone was iced up and freezing. We camped in a grove of black spruce out of the wind, lit a huge fire and melted the ice buildup on the skidoo, as well on ourselves.

Back in the MLDC boardroom, we found it difficult to count the annual rings on those cookies. Forest tent caterpillar attacks trembling aspen every so many years. It eats the young leaves. The tree growth slows down and the growth rings barely show up. The tree seems to stop growing. And, without leaves it could not grow, of course. I told the boys that trees knew when there was a bad chief on the reserve because they ceased growing.

Allen Bird accepted a new windshield with grace and patience.

Montreal Lake Development Corporation presented the report, based on samples from 132 trees, to a consultant representing the chopstick developers. He used the MLDC data and pretended that it was his. Then to cover his tracks, he bad-mouthed Silviba. If it was not so ugly an incident, perhaps the chopstick factory could have succeeded. The crew from Montreal Lake did excellent work from fieldwork through to the wood analysis.

There is a story behind the games Silviba played to get its $10,000 fee for service (expenses included). Henry Naytowhow was chief. I had bought a set of cassette tapes covering the subject of 'negotiating'. I left Pelican Narrows at 4.30 am on a Wednesday to take the back road into MLDC at Montreal Lake. All the way there, I played the tapes. Indians are reluctant to pass money over to non-band members. To reduce this risk, I had contacted the federal agency responsible for the project funding and asked that it make the cheque out to both Silviba and the band. When I entered the MLDC boardroom, I found the chief with Alphonse and Norman Ross. Several routine matters came up for discussion, including my plea for the band to hire a forestry technologist to grow the band's forestry business. I acted as if the cheque was the last thing on my mind. However, I considered all the disasters about to occur if I walked out of the boardroom without the cheque. The meeting ended, and as I went through the door to the hallway, Chief Henry said: "Oh Tony, I have a cheque for you". He took it out of his back pocket and passed it over. In time, Chief Henry taught me some Cree, and Norman Ross invented a new name for me: 'Mr. Hungry'. In Cree this sounds like 'nahtikatan'.

The Hiring of Eugene Kimbley
After Weyerhaeuser bought the pulpmill and surrounding sawmills in 1986, Montreal Lake Cree Nation asked me to look for a suitable director of forestry. The Weyerhaeuser Forest Management Agreement area was part of the band's traditional landbase. The provincial government had not consulted with the Cree Nation beforehand. Remember? I warned the Prince Albert District Chiefs through its economic development director that this may happen. It did.

I remember a later meeting with Chief Henry Naytowhow and councilors Norman Ross and Alphonse Bird. We discussed again the intension to hire a director of forestry for MLDC. When Alphonse raised Gene Kimbley's name as a possible candidate, I responded that I was searching for a candidate with a strong technical skill set. I did not believe that Gene was suitable. After that comment, both Norman Ross and Alphonse became really angry, almost foaming at the mouth. Norman told me that he had shot out the cherry on a police car, which was a hint that he could use me as a target just as easily. They stormed out of the boardroom and Gene Kimbley became the MLCN director of forestry. Geno had started out as a resource officer in the little hamlet of Armit, at the north end of the Woody Lake Road.

Fascinating. Gene and I became a team that lasted at least thirteen years. Geno was Métis, and was raised in Beauval. Like most of those living on the northwest side of Saskatchewan, he was Indian in looks; although his great-great-grandfather could have worked for the North West Company. He was sent away to school at the Battlefords, excelled in hockey, to the point that he became an NHL prospect. I used Geno, who spoke Cree, to get inside the Indian world as means to generating First Nations forestry initiatives, and Geno used me as his technical resource. Over the years, we became 'hand and glove'.

Geno had come away from his hockey years with a severe alcohol problem. Later, this got worse. One day he was returning home on Highway 3, somewhere west of Shellbrook. Inside his vehicle, all by himself, he received a life altering message and committed his life to Jesus right then and there. 'He got religion' – Hallelujah. I was too busy to attend church, but Geno used to make sure that I had a ticket to the evening supper meetings of the Prince Albert chapter of Full Gospel Businessmen's' Association. He was a national director when he and I joined forces. These occasions were a wonderful circle of fellowship and worship after a full tummy. People would travel from all over to present their testimony. It was in this atmosphere, after one night's meeting, when I walked to the front and became "born again". I could just feel the pull. An incredible once-in-a-lifetime happening. – Hallelujah. Geno said that I was never the same after that.

Geno had an astute political sense. I used to write his speeches on several occasions. He became a spokesperson for the 'Indian' side at meetings such as the annual meeting of the Canadian Institute of Forestry held in Banff one year. Gene was a complex person, hugely smart, and he could play the racist game when he felt threatened, or when it served a strategic purpose. He displayed hostile tendencies as well, probably left over from his hockey days.

This was not all of my speech writing efforts. I wrote speeches for Chief Noland Henderson of MLCN, and campaign literature for Henry Morin, a band councilor at Pelican Narrows, at band election times. Henry did not make it on the last occasion.

All was not 'wine and roses' for Geno at Montreal Lake Cree Nation. A few individuals treated him as a slave. To emphasise this situation, he was sent up to Halfway House, a band restaurant some 40 km up the road, to deliver cigarettes and other supplies regularly. On occasion, Treaty Indians consider people of Métis descent to be 'slaves'. He commuted from Prince Albert each day in his own transport. The cigarette supplier also lived there, so possibly this served as a rational solution for the Halfway House cigarette supply dilemma.

Silviba Services advised, mentored and trained band members through Gene to undertake all phases of forestry fieldwork over the next few years. Gene successfully negotiated for a small operator annual timber permit for the band, supervised tree planting and tree thinning on Weyerhaeuser lands and set up a sawmill on reserve.

Remarkably, Geno used the Silviba (Frank Hart)-generated five-year forestry business plan template as his bible.

Montreal Lake Band Enterprises

Geno developed the band's forestry business. Until then, it was an orphan run by Christian Nelson, band councilor and Jim Bradfield, a Métis foreman who had taken possession of the entire band's equipment – log skidder and D8 cat. Geno rounded up band money to

190

replace the final drive on the cat, as well as operators who could keep the wheels turning without causing too much equipment damage. This was difficult politics for Geno – too many cooks spoiling the broth. I mentored Geno in the acquisition of an annual volume of adjacent Crown timber. Wilson Bird was the band tree planting foreman. There is a photo of him planting his millionth tree in a Weyerhaeuser timber harvest area. Montreal Lake tree planters had developed a good reputation for this by then.

Shareholder (38%) of Silviba Services Ltd

In 1991, MLCN became shareholders in Silviba Services Ltd. It was given 38% of the Company. I held the remaining 62%. The band provided $9,000 towards the purchase of specialized GIS (Geographic Information Systems hardware and software). Also, band members would take part in GIS training on the job. Longer-term opportunities could lead to a growing capacity in co-managing the traditional forest landbase with Weyerhaeuser Canada. Silviba nurtured Delilah Bird in her training as a GIS tech. But, this was not to last. First, she had a problem with transportation needed to get her to Silviba's office in Prince Albert from her home in Montreal Lake. Second, Noland Henderson had just quit his school principal's job on reserve to become chief. He and Delilah set up a sustainable environment management unit within Montreal Lake Development Corporation. To me, this was akin to a tugboat that had lost its radar, and was stuck in a fog bank on the Salish Sea.

Chief Noland Henderson became a SaskPower director shortly thereafter, as well as his more usual duties. His wife was of Ukrainian descent and kept an eye on the band administration. This was after band funds had vanished when the band administrator was caught stuffing away band funds into his own account.

At a supper meeting in late 1997 attended by Montreal Lake Cree Nation council and Chief Henry Naytowhow, I briefed them on Silviba's failing financial position. I told them that Silviba's digital mapping capacity was out-of-date and that I could not bid on further Weyerhaeuser or government work. I told the chief that I was buying out the band for $1 and possibly could face bankruptcy. I put a loonie on the table and left. The band then went to its lawyer to bring me to

191

court for fraud. The lawyer, who knew me to some extent, defended me and explained that it would all be a waste of time and money.

MLCN Sawmill, 1991

Figure 23: MLCN erected a Mobile dimension sawmill similar to this in the Development corporation's yard.

Montreal Lake Development Corporation bought a Mobile Dimension sawmill manufactured in Oregon. I arranged the purchase and Gene went down to the factory and towed it home. On the basis of our business plan, six full time jobs for band members on reserve would evolve. Marketing was no problem. The peddlers never had enough supplies on their trips south and the band families would buy all the lumber available. The logs came from the reserve. Gene arranged for his boys to build a good lean-to as weather protection. After some feeble efforts to make lumber, the project faded away. I asked Geno several times why his project collapsed, but never did get a satisfactory answer. Eventually, the development group sold the mill to a band up north. The best conclusion that I could figure out was the band considered this little mill too small to suit its dignity.

Jackie Ross, logger (1)

Jackie Ross and his dad, Sandy, together with Buddy Forrest and Jackie's brother, were Montreal Lake band members, but lived off- reserve. Sandy lived on his trapline at Two Forks, on the way to La Ronge. He had run the preliminary centreline for Highway 2 North from Timber Cove to Weyakwin before this highway was built. When I first knew Jackie, he was living in Paddockwood. This family had started out in the bush as part of the logging training undertaken by the woodlands management of Parsons and Whittemore of New York, the builders of the Prince Albert pulpmill in 1965-66. The Ross family was tireless. Whether it was mowing down fencepost timber, pulpwood or sawlogs, they were top producers. Jackie and I worked closely connected on several forestry projects. However, this relationship matured when I travelled all over with Jackie to scout out potential spruce sawlog supplies from private landowners. I learned to count the number of loads of logs expected from logging a forest parcel.

Jackie did most of the logging of the prime white spruce timber on reserve at Montreal Lake. The poplar went to Weyerhaeuser in Prince Albert and the white spruce to Tolko at The Pas. Later, he salvaged timber off a forest fire just south of the community. He had his share of tripping over bureaucracies, none of which he had time for. Silviba did the cutplans and timber volume calculations.

In later years, when Lionel Bird was chief, Jackie's contracting work for Montreal Lake Band Enterprises earned the band big bucks. But, when Jackie needed money for equipment acquisitions and repairs, he found that chief had emptied the Enterprises bank account. Jackie retreated to Two Forks and lives there to this day. He is hired up north, or does posts for Aallcann Wood Suppliers Inc. in Prince Albert.

Of course, there could be mitigating circumstances. Perhaps Indian and Northern Affairs Canada (INAC) scooped the bank account to balance its books. However, it is difficult to rationalise the closure of a successful enterprise that made money.

Ω

Chapter 18:

Peter Ballantyne Cree Nation (PBCN) forestry

I have just talked with Chief Peter A. Beatty in the band office in Prince Albert. I have permission to include the following:

Extract from the Peter Ballantyne Cree Nation website:

> *"The Peter Ballantyne Cree Nation people have occupied lands in northeast Saskatchewan since time immemorial. There are no legends of the Cree coming into the area and wresting the lands away from another First Nation. The roots in the area run loud and deep. Pottery shards found in Northern Saskatchewan have been dated to about 1000 years before present, and indicates without a doubt that they were particular to the Rocky Cree. Although it establishes that the Rocky Cree have been around a long time, no archaeologist can be certain for how long. Paleoindians were in the area shortly after the retreat of glaciers and a projectile point found near Southend has been dated to 5,600 years before present."*

[I beg to differ: I recognised certain physiognomy that identified with Sioux Indians. My friend, Laurence, a Deschambault resident responsible for the log home building training program, indicated that his ancestors could have been Sioux, given that Sioux warriors came north on raiding parties – TR.]

"Traditional Territory: *The area traditionally occupied by the Cree Nation encompasses about 20,000 square miles, from the Saskatchewan/ Manitoba border west to the west end of Trade Lake, north to Reindeer Lake, and south to Sturgeon Landing. This was the traditional hunting/gathering area of the Cree Nation.*

The Rocky Cree*: The Peter Ballantyne Cree Nation is called Assin'skowitiniwak or Rocky Cree. Assin'skowitiniwak means "people of the rocky area". This rocky area is the Precambrian Shield of northern Saskatchewan. The Precambrian Shield is that band of land between the southern parkland and the northern tundra. The terrain is primarily muskeg, lakes and thin soil with much of the granite laid bare by glacial action. The forest cover consists mainly of white and black spruce, jackpine, trembling aspen, and some birch. It was in this rugged environment that the Cree Nation's ancestors made their livelihood, hunting the much prized moose and other game for their sustenance.*

Mode of Living Prior to Fur Trade*: Their primary transportation was the birch bark canoe and they traveled their country through innumerable Waterways which they thoroughly knew. They were relatively undisturbed even by other First Nations until the arrival of European traders at Hudson Bay beginning around 1680. Prior to that, they had lived on a subsistence basis, taking from their environment only what was necessary for their survival. They did not trap for fur except for what was necessary. At this time, the beaver would have hunted primarily for its good eating.*

The Cree Language*: Cree is spoken in the "th" or Rocky Cree dialect. This differentiates it from the "y" dialect of the Plains Cree and the "if" dialect of the Swampy Cree. Historically, the "th" dialect was spoken as far south as Cumberland House, which now has the Swampy dialect. The Rocky Cree group extends as far west and south as Lac La Ronge and Montreal Lake and as far north and east as Brochet, Southern Indian Lake, and Nelson House in Manitoba.*

The Fur Trade: *As stated, trade developed with the Hudson Bay Company shortly after 1680. A brisk trade developed and the Rocky Cree became trappers for the first time in their long history. Furs were traded to the Hudson Bay Company for goods such as rifles and ammunition, chisels, axes, blankets, pots, and other commodities. At* *times they may have traveled to York Factory on Hudson Bay, which would have been about a one-way 15 day trip from Pelican Narrows".[38]The fur trade came to the doors of the Rocky Cree in 1774-75 when the forerunners of the NorthWest Company traveled through Amisk Lake, Pelican Narrows and Trade Lake, seeking to expand their trade with First Nations further north and west. In 1776, the Frobisher brothers and Alexander Henry the Elder established a short- lived trading post (Forte du Trait) at Frog Portage (Portage du Trait) on Trade Lake."[39]*

Peter Ballantyne Cree Nation occupies lands within treaty 10, 6, and 5[40]according to the map.

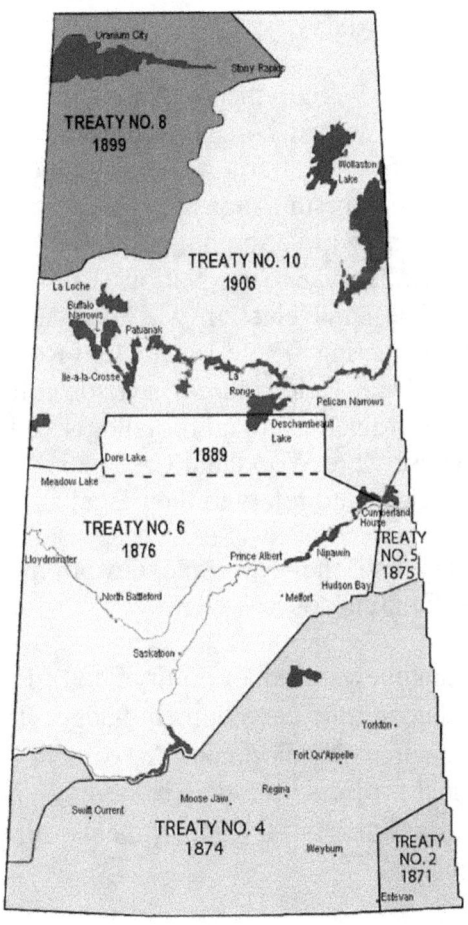

Figure 24: Indian Treaties, Saskatchewan, otc.ca

197

In all the years that I took on the role of band forester, never once was I a party to discussions on the complexity of the treaty arrangements within the band. I assumed that for all intents and purposes they were Treaty 6 people.

Peter Ballantyne Cree Nation, Pelican Narrows (PBCN)

On my arrival back home in Regina from a teaching stint at Lakehead University, I prepared the family for a move to Prince Albert. The kids elected to stay in Regina. That is what happens when a family gets split up. For the first while, I commuted. On Nov 30, 1985 we became grandparents to Matthew, and Marilyn stayed in Regina to help care for him. She finally joined me in Prince Albert after Matthew's first birthday.

My first office was in the Valhalla apartments on Branion Drive. I put together two sawhorses supporting a sheet of plywood. However, I felt out of touch after a while. I do not remember how this came about, but one day in August I was invited to sit down in someone's boardroom with a brand new chief and council. They represented Peter Ballantyne Cree Nation (PBCN) headquartered at Pelican Narrows. The band election was over and this was the first initial band council meeting. There I met Chief Ron Michel, councilor Henry Morin, and Dale Reid, chief's advisor/gofer/problem solver. Dale was university-trained as an anthropologist, a brilliant individual and a survivor. These were Woodland Cree people and lived in forest surroundings. Their website refers to their label as Rocky Cree.

Henry Morin had the economic development portfolio[41] for Pelican Narrows. Funding sources were:

- SIEF (Saskatchewan Indian Equity Fund), Prince Albert
- Band level economic development budget (Pelican Narrows)[42]
- Forest Resource Development Agreement funding for forest management projects on Indian reserves and private lands.
- Aboriginal Business Development Program (ABDP), Saskatoon.

In the winter of 1985-86, I received approval from the Forest Resource Development office in Prince Albert and the band to commence forest mapping and inventory of Peter Ballantyne reserves:

Powerline Clearing (1)

Chief Ron Michel was the new chief, gung ho and ready for action. Dale Reid had discovered that SaskPower was letting out contracts for brushing powerline rights-of-way along highway 135 from the Hansen Lake road junction to Pelican Narrows. And from the junction west along Highway 6 to Deschambault, a total distance of some 46 km. Silviba helped him put the bid together, then ribboned out the right-of-way boundaries and supervised the operation. SaskPower had the option of hiring mechanical brush-clearing contractors, but stayed with the band's bid.

On an early spring morning in 1986, snow still around, Chief Ronnie emptied Pelican Narrows. One hundred treaty Indians showed up to work. Between everybody, they had six chainsaws operated by trappers. Everyone else equipped themselves with axes. All but a few walked to work. On the second day, they were down to sixty workers, and by the start of the second week, twenty or so showed up.

By this time, Silviba had hired a forest technician, and PBCN took possession of a Timberjack log skidder.

Figure 25: This resembles the Timberjack log skidder operated by Henry Morin on the powerline clearing in 1986. (https://www.google.ca/search?q=timberjack+log+skidder).

Saskatchewan wildlife staff provided strict rules on placement of brush created as the clearing progressed. Henry used the skidder to push the brush piles away from the timber edge along the cleared right-of-way as instructed by the resource officer at Creighton. He was also instructed to leave openings in the slash for wildlife travel. Eventually, the band completed the project, was paid, and Silviba received its fees. I served as the liaison between SaskPower and the band. I was also well aware of SaskPower's 'soft side' with this contract. The Corporation could have hired contractors from down south equipped with mechanical brushing machines, but elected to give the manual work to Peter Ballantyne Indians. Mind you, the rocky ground could have done much damage to mechanical cutting heads.

Nikotawsik Forestry

Dale Reid registered the band's first forestry company: Nikotawsik Forestry something or other. Everyone received a hat. In 1986, Dale and I selected a six-member forestry crew from the young men in Pelican Narrows. I taught them to map and cruise the timber on reserve. It was old and just waiting for the next forest fire.

I discovered a retired sawmill instructor from the Indian agency in Shellbrook, Stan Read. Stan had found a portable sawmill, overhauled it, and came across a farm tractor with a power take-off. Nikotawsik took possession of these items and moved them to the sawmill site in Pelican Narrows. We taught the boys the workings of a Jonsered chainsaw and how to fall timber safely, then skid it to the sawmill site with the band skidder. They had all the correct safety apparel. Dale got orders for timbers required by the new Pelican Narrows school construction and Stan, with the crew, sawed up the logs to the specifications needed. Later, in the spring, local residents put in orders for fence lumber and dock timbers.

But, what started out so well, fizzled to a complete stop. Although, Henry Morin as councilor in charge of economic development, was also in charge of the sawmill's cashflow, he was never in the community to track progress. Thus, in the new fiscal year (1987) the training money ran out, and this excellent project collapsed. Another heartbreaker.

200

Henry became known as '*Mr. Per Diem*' since his economic development funds allowed him to travel all over the place. He got home Fridays. It is a journey of 4.5 hours from Pelican Narrows to Prince Albert. The road south from Pelican Narrows, only built in the early 1970s', was winding, the gravel in places was blasted rock. If travelers did not lose a life, they got a flat tire. In all the years I hung around Peter Ballantyne, I never stopped worrying about them on that road to Prince Albert – summer or winter, or snowstorm or wind. Drivers could run into elk on The Narrow Hills, skid off the road in black ice, and as for moose: I almost hit a huge moose charging into the road on one November night just south of Cariboo creek. I was towing a skidoo trailer, and that is what almost tangled with the animal. Before roads, everyone travelled by boat or canoe to the nearest point of road access, or up through Frog Portage on Trade Lake, part of the Churchill River system.

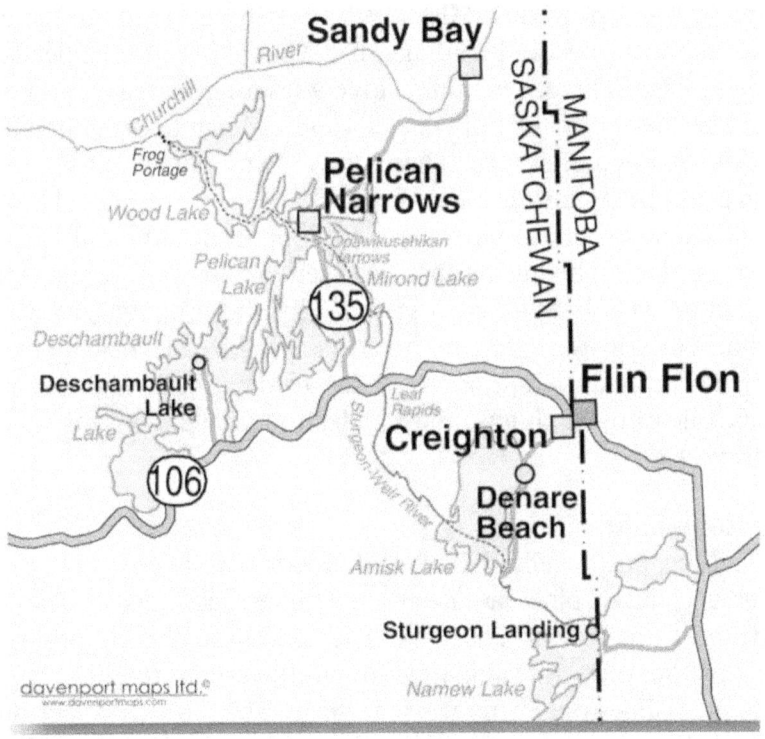

Figure 26: Canoe route from Pelican Narrows, north to Frog Portage on the Churchill River. Courtesy of Davenport Maps, Victoria, BC©

This was Alexander Mackenzie's turf, North West Company, back in the late 1770s. Henry Morin took me out to fish for pickerel on Pelican Lake one time, and then travelled up the old canoe route to show me the petroglyphs at Medicine Rapids.

Opawikusehikan Narrows ties Pelican Lake to the very top end of Mirond Lake. This was a major transportation route in the 1770-1870 era for the fur trade between Frog Portage on the east end of Trade Lake on the Churchill River and Sturgeon Landing to the south. It was a great place to hijack traders travelling up and down the Sturgeon-Weir River.

In the spring of 1987, I put on 4,600 km travelling between Prince Albert and Pelican Narrows. By this time, I referred to myself as the band forester and stayed in touch with Chief and council. By getting up at 0400 hours in Prince Albert, I arrived in Pelican Narrows at 0830 hours. I parked at Mistanosayew fishing lodge and cafe, run by the band just outside the village, had a snooze, and then headed for the band office. There, I would receive a briefing on the day's council agenda, get permission from Dale or Henry Morin to bring up an item that needed council's OK. There were two matters afoot: timber cruising and mapping the other Peter Ballantyne reserves. However, chief and council could never lose the opportunity to put the white man in his place. This meant that I had ten minutes before the meeting adjourned, normally at about 6.15 pm. This did not interfere with my mission, which was to get the band involved in the management of forestlands around them, nor to my day-to-day relations in general. It was just one of those things, and I understood the underlying rationale for this perfectly.

Limestone Lake

During one winter, Nikotawsik got a permit to log a patch of overmature jack pine adjacent to Limestone Lake. The boys commuted from Pelican Narrows. They accessed the logging site by driving off the road, through a gravel pit, then on to the ice of a small lake beside Limestone. This was February. The ice was 4 feet thick – excellent. We planned to finish logging and hauling by the middle of March to avoid the spring thaw. The boys, though, did not manage to skid that wood onto the ice until April. Pelican Narrows had hosted two winter sport weekends in the meantime. So, between parties the boys failed to show up. Rod Morin operated the skidder. Each time he piled up the jack pine on the ice the bark exfoliated, until he had covered the entire decking area in a layer of bark. I learned much the day Rod put the skidder through the ice[43] – he barely got out of the cab in time. The shredded bark created a hot spot that melted the ice underneath. While salvaging that skidder, an enormously complicated job, the salvager cut the ice into blocks with his chainsaw, so he could attach a winch line to the machine. He was from Creighton, and the previous owner of the log skidder. Those blocks of ice were honeycombed underneath. Why? Ice melts from below. Nikotawsik had to release the skidder, as was, on the lake floor, to the salvager as payment. Otherwise, Nikotawsik exposed itself to a hefty fine for polluting the lake. That skidder started a new life as a diamond drill rig in the region around Creighton. The salvager also rescued the logs and trucked them into Creighton.

Limestone Lake was one of my sleepover locations on my Pelican Narrows trips. An old road ran down beside the lake. At the end, it turned into a cleared area, with a set of stairs going down to a float dock. I slept in the back of the old green suburban. The cacophony caused by waking birds of all descriptions and the bombardment from squirrels dropping pine cones onto the truck roof, made for some unforgettable mornings. I loved that camp spot.

Birch Portage reserve

Isiah Custer was the trapper at Birch Portage. Silviba and the band's forestry crew stayed at his cabin, and Isiah cooked, while the crew mapped and timber cruised the reserve. Travel was by skidoo and snowshoes; it was winter, and Isiah set fishnets through the ice. One evening, after returning to camp, we ate a meal of the finest fish I had ever tasted – Mariah. Seemingly, it was only at its best in the winter season.

I was entranced with Isiah's knowledge of all the elements that contributed to his surroundings. Without maps or compass, he just knew where he was at any time, day or night. He knew his trees, muskegs, geology (rock outcrops), ice thickness, animal populations. All in his own Cree language, of course. One time, soon after, on a warm, damp January day, Henry, Isiah and I ribboned out an access road linking the reserve to the main road into Creighton. Isiah was in the lead. The visibility was poor, due to drizzle, low light and low cloud. At one point, Isiah navigated between a large rock outcrop on one side and a muskeg on the other. Neither Henry nor I were aware of these as we followed Isiah out to the road.

Birch Portage reserve, with its new road access and bridge across Granite creek installed by Stan Read, became the next forestry focus for the Indians at Pelican Narrows. Dale Reid registered another company, Mirond Lake Forest Products Ltd, to undertake logging and sawmilling at Birch Portage now the road was in. One spring day, Henry, Isiah and I, and others walked along a trail from Granite creek to Birch Portage – It was along the old portage. Henry found a nest of grouse eggs, which we boiled and ate for lunch. The eggs were scrumptious. Isiah told me the grouse parents would replace the eggs in the nest. While in that vicinity also, Henry found some rat root growing close to the creek. Indians used this for coughs and bronchitis, according to Henry. He told me to take some home.

As an aside, one February day in pure sunshine, Henry took me up to the north end of Granite Lake. I wanted to check on the quantity and quality of spruce there. Henry had thrown a sack of caribou ribs on to the skidoo for use as a lunch. These bones travelled south from just inside the NWT north of Brochet at the top of Reindeer Lake, down to Southend. Henry's Southend relatives made sure his family in Pelican Narrows was not left out.

The white spruce timber logged at Birch Portage was of premium quality. There was a bunkhouse placed at the sawmill site. The sawyer lived in Creighton, one hour east. As well, the band trucked peeler logs off the Birch Portage reserve over Stan's new bridge, to mills in Hudson Bay. Under the Canadian Forest Service mandate, the logged area was scarified[44] and planted with replacement trees. Eventually, Mirond Lake Forest Products crashed because of equipment breakdowns and management. Henry built a house at Birch Portage. It became his family's weekend escape.

Mirond Lake and Amisk reserves

Silviba and the six forestry crew members from Pelican Narrows undertook the timber cruising of these two reserves in due course. Both were water access. The crew camped on beaches in both instances. This was historical land. The timber on the Amisk reserve was logged back at the start of World War II on the orders of the federal government. As of 1987, the band had received no compensation. I believed that this was an excuse by local operators to steal the logs at the time.

Arthur Ballantyne and the forestry crew

Arthur Ballantyne was, and we hope still is, a resident of Pelican Narrows. He had family in Thompson, Manitoba. Art loved to read any book by Louis L'Amour. If Art ever reads this, he was known as the ugliest Indian in Pelican Narrows. However, I say this with love in my heart. Arthur Ballantyne was the best forest surveyor I ever had the pleasure to work with. On that Amisk job, he ran a 4,000-metre baseline with just a Sylva hand compass. The swing error was 4 metres. Incredible. I knew how well he had done. We tied into a rock hump that showed up on the air photo, and that is how we calculated the 4 metres. Art was hired each winter to run lines used by diamond drillers to drill test holes. He had a job there whenever he needed one. In 1988, I took Art with me on a Silviba contract to British Columbia. But more on that later.

Those boys were the cream of the crop. They may have stayed in bed a little longer, but I could not get them to return to camp until after dark. Back on that first job at Pelican Narrows, we toured the forest and learned the names of all the tree species we would come across during our daily work. They were familiar with all the tree species, but in Cree. Then we went over the procedures for mapping the terrain and taking tree measurements – species, total height, diameter at breast height, and finally, total age. In addition, during the mapping we had to draw in the boundaries as we crossed from one tree type (ecotype) to another. One evening, on their way back to camp after a hot, sweaty day's work, the boys discovered moose tracks crossing Art's baseline. That was it. I ate by myself that night. I do not know what they would have done if they had caught up to it. We had no guns in camp.

Art and I had our own special job at Amisk. We had to run the east boundary for the reserve. It was swampy, full of alder brush, and miles long. At one point, Art told me that 'I was devious'. I think I had just talked him into running the line across a piece of wetland. And, believe it or not, we got lost. At the end of the job, we were to find the boat parked at the lake edge in a certain spot. The shoreland was wetland, the lake level almost at shore height, and the brush thick. We had no vantage point to use as a reference. In addition, the shoreline changed direction constantly. We had the air photos and the compass, but no point of reference. I witnessed a Cree Indian who was lost! Our predicament was not to last long. We stumbled on to the boat eventually. I give credit to Art for finding it. It was August – hot and sweaty.

The Amisk job meant much to me. It was at the south end of the Sturgeon-Weir River system. Back in the fur trade days, it linked the Saskatchewan and Churchill Rivers. Dale Reid showed me David Thompson's diary of his winter journey up the Sturgeon-Weir River. He recorded a visit to a Cree winter camp. The parents had just killed and eaten their little boy to stay alive. On a trip up the River, I found a community graveyard perched on a slope above the bank in the middle of a wilderness. It was so peaceful.

Before I exit Amisk reserve, I need to mention the crew found the tallest white spruce ever recorded in the region, 34 metres plus. Later I took Mike Newman, the Canadian Forest Service Prince Albert director, up to that tree to confirm our measurements. He agreed. A sixteen-hour day on that trip.

With reference to the Amisk map displayed earlier, the Sturgeon-Weir River flows south from its source at Pelican Lake. It starts at the Pelican Narrows rapids, then proceeds southeast through a number of lakes, before flowing into Amisk Lake. From there it proceeds south to Sturgeon Landing, its southern terminus on Namew Lake. The Sturgeon-Weir River was the main link for explorers and fur traders between the Saskatchewan and the Churchill Rivers. Silviba and crew could have plausibly set up the Amisk camp on the same spot as David Thompson. In fact, they must have since there was no other accessible site.

Peter Ballantyne Cree Nation Integrated Resource Management Agreement Proposal, 1988

I presented a draft proposal to the band. A comanagement renewable resources agreement on provincial Crown lands. It covered the traditional landbase recognised by the Peter Ballantyne Cree Nation (PBCN). The purpose of this agreement was to:

- Create long-run employment for band members
- Develop industrial partnerships that would create cashflows to expand business opportunities
- Provide a sense of responsibility for sustainable management of the lands in question
- Create small business opportunities and professional development within the integrated renewable resource management agreement

I presented this outline to Chief and council one day in May. Dale Reid set up a meeting with Minister Lorne Kopelchuk in Regina. The meeting went well. Then, two more flies landed in the ointment. First labelled Jack Falk, and second, the need for a letter from the provincial government committing to a timber allocation for the band sawmill at Deschambault. The latter became a condition of the forthcoming federal 'Resource Access Negotiation' application in support of the proposed Integrated Resource Management Agreement.

Jack Falk was the recycling/garbage agent in Creighton. He got wind of Peter Ballantyne discussions about the formation of a Forest Management Licence Agreement application (his translation), and got to government officials first. A 'due diligence' investigation by the band suggested that Mr. Falk had no prior experience in the forest industry at the level required. However, he had access to technical support when needed. The bureaucrats in Regina considered Jack Falk more acceptable than the band, at least in the eyes of Alan Appleby, assistant deputy minister of Natural Resources in Regina[45]. But, the Peter Ballantyne Cree Nation still needed a timber quota with which to support its RAN application. Band politics locally was a great part of the pressure to persevere.

Peter Ballantyne Cree Nation – Deschambault Lake

I dedicate this section to the memory of Oscar Beatty and Jack Custer.

Where to begin:

In his own way, Jack Custer, who represented Deschambault Lake on the PBCN council, was a 'go getter'. This community was two hours closer to Prince Albert than Pelican Narrows, but did most of its business in Creighton or Flin Flon. It also sequestered a sawmill set up back in the Department of Northern Saskatchewan era. I first met Jack when his boys were logging fenceposts south of the reserve. Jack was running a Timberjack forwarder in the middle of a big pile of logging slash. Perhaps Jack had met me at a council meeting in Pelican Narrows, and then invited me over. Sounds plausible.

The projects that I helped Jack with over the next few years were complex and ambitious.

The Sawmill

An old sawyer left over from the DNS days ran the mill after Jack moved in a bunkhouse and kitchen unit to the sawmill site. It did not take long before local youth smashed it, but the sawyer persevered and had trained a moderately focused crew to make lumber. Jack had a standing order with a lumber broker in Saskatoon who came up with a flat deck truck when there was a load of lumber ready to pick up. Occasionally, Jack had sold it to someone else, and trucker was faced with an empty yard and a wasted expense.

A sawmill operation needs a log supply, so me or Stan, or both, found Jack a log skidder and Stan trained the crew. The band bought a grapple loader to lift logs on to the sawmill carriage. It bought a forklift to move the bundled lumber into the yard for storage, or onto a truck for transporting away. I did the logging planning and coordinating with both Indian Affairs and Forestry Canada. Good timber surrounded the sawmill site. But at certain times of the year it was too muddy to carry on.

Absenteeism plagued that operation, no money for payroll, and sometimes, no market for lumber, and probably no logs to saw. Through it all, Jack kept plugging along. Finally, a sawyer trainee rammed a log into the saw and bent the mandrel. After that, Jack concentrated on logging. I did the paperwork, produced the cutplans, did the timber reconnaissance with our Deschambault forestry crew, and liaised with Indian Affairs and Forestry Canada. Amongst all this, I made sure that reforestation followed logging. By this time, Indian Affairs had a forester of its own, and Canadian Forest Service had its own field technician. Once more, I was the liaison.

On one occasion, Henry Morin was in Deschambault at the same time as me. For some reason, Jack Custer, Henry and I held an informal meeting in the kitchen of the community hall early in the afternoon. Henry was pissed off. He and Jack conversed in Cree, meantime glaring at me. This was not their usual behaviour. I was never given an explanation, but could only conclude that something I had done jeopardized Henry's *per diems*. Sometime later, Henry asked me to do a proposal for his son, Terry, who wanted to go to work log skidding. I did several of these over the years for band members.

Logging

Jack got Jackie Ross to log the balance of the accessible timber on reserve, then delivered this to Tolko Industries (pulpmill and sawmill) at The Pas. More of Jackie later. This triggered the need for a comprehensive forest management plan covering a longer-term window. Silviba got the job through the band's director of economic development. Meanwhile, Jack Custer cranked up logging capacity at Deschambault. With Dale Reid's help, he found the money for a mechanical logging operation. This comprised a feller-buncher, grapple skidder and log loader. Log production was trucked to Tolko at The Pas. The equipment was operated by band members, with some coaching from Tolko woods supervisors. I mentored Dale, and he, in turn, mentored Jack.

The evolution of forest management planning at Deschambault

I saw this as more than a forestry-based entity. It would generate jobs in the management of fish, animals, tourism and forest fire. I saw this as an Integrated Resource Management Program. This would call for a contract between the province and the band. Chief and council did get to visit Honourable Darrel Cunningham, the Natural Resources minister in the new NDP government, later. However, the bureaucratic roadblocks remained in place.

The area considered for a comprehensive forest management plan covered the traditional landbase "occupied" by PBCN at Deschambault. This was the landbase supporting the band hunting and fishing activities, and traplines. I recall one of the final meetings in the Deschambault boardroom. I had the numbers and a plan on maps of the projected logging locations for the next five years. Over the years, I had accompanied band members and Jack on various expeditions that provided intelligence on timber location, both quantity and quality. Access was also of significant consideration. Some logs would have to be trucked using ice roads. For instance, the West Arm of Deschambault Lake contained a big volume of large white spruce. It needed salvaging. Silviba and crew had used an increment borer to determine total age and found that the wood was turning pink. This meant incipient decay because of age. Jack Custer and I had gone fishing up on the West Arm one day, and first saw that spruce. On another occasion, the band council had its meeting in Deschambault. Chief Ronnie and the boys took me with them up the East Arm of Deschambault Lake. They never let me forget the jackfish I hooked through the anus.

Nature uses forest fire as a cleansing agent and forest replacement tool. Much of the forests around Deschambault were overmature, dry and subject to large fires from lightning. I warned the community that eventually there would be a "Tony" fire in their vicinity. This condition influenced my judgement as I formulated the five-year forest management plan, including proposed road access. I never ceased to impress on the Deschambault community that it was surrounded by overage forests, very dry after a long period of drought. Ripe picking for a thunder storm with lightning.

210

Thus, it became Oscar Beatty's job to provide an overall perspective on the potential impact to the landscapes in question. He was the vintage elder, knew the country like the back of his hand, knew the animals and spiritual places, and served as the Deschambault 'Wikipedia'. His winter home was on Sandy Lake, at the west end of the plan area. Silviba had a full-scale GIS (digital spatial planning) capacity in-house by this time. This created the opportunity for a professionally packaged forest management plan thanks to Heather Patterson, Silviba's GIS operator.

Jack had called the meeting in the Deschambault band office boardroom. The maps were on the wall. Oscar sat across from me. He was distant and aloof. Folks were going in and out to attend to other business. Finally, Jack as chair, brought the meeting to order. Peter Beatty, a fellow councilor (now chief), sought to determine how the band would benefit if the plan was approved. From then on, the discussion loosened up and Oscar took his turn. His 'put down' attitude was directed at me. I was a slave. Then everyone got up to look at the maps and refill their coffee cups. I looked over at Oscar and grinned. He winked back. He had only behaved as he was expected to. The meeting adjourned with no follow up instructions except that Dale Reid and Jack Custer were to contact Tolko and Carrier Lumber and explore log delivery opportunities.

Plans for Deschambault sawmill no.2
I had worked with another client who had engaged a sawmill design expert to plan for a new sawmill. In 1994, Deschambault forestry, through Ron Ray, the band's economic development councilor and Jack Custer, paid for a mill design compatible with the timber profile found on the traditional lands within the proposed forest management plan. This would have created 18 local jobs and supplied the market demand for lumber all the way to Flin Flon in Manitoba, then south to The Pas. However, the project capitalization was too large for Deschambault. Jack Custer took the project business plan to Peter Ballantyne Cree Nation head office, now located in Prince Albert. The economic development folk there were into big time projects such as a casino and a motel, among others. So, white guys working for PBCN economic development kept this forestry initiative to themselves and left me out of the equation. A hugely

211

viable project died 'on the vine' because of a lack of political will and 'turf'.

$$\Omega$$

Before proceeding, I have to stop to tell a story. There is a road junction just west of Big Sandy. Highway 6 (the Hansen Lake road) proceeds north from Smeaton. It travels past Cariboo Creek Lodge at mile 46, then past Bear Lake Lodge. It then proceeds up through the Narrow Hills Provincial Park before turning sharply to the east at this junction. Highway 165 leads from the junction west to La Ronge. Highway 6 continues east to Creighton, Saskatchewan and Flin Flon, Manitoba. Peter Ballantyne communities at Deschambault Lake and Pelican Narrows use this highway to scoot back and forth to and from Prince Albert and Saskatoon. Highway 165 is gravel, while Highway 6 is now paved, but was once a gravel road.

In 1985, when I first drove up from Prince Albert, the highway east of the 165 junction passed across a long stretch of muskeg. In the spring, and again in the fall before freeze up, it was impassable. It was upgraded by loading down the muskeg portion with a thick blanket of sand. For the next three years the sand gradually pushed out the soggy, mushy muskeg, Saskatchewan Highways department finally had a road grade resting on solid rock. After they paved the new grade, the only bumps present were those created by culvert pipes that still lay on the muskeg subgrade. Anyway, the speed limit went up and everyone cut a half hour off the travel time to Prince Albert. This straight stretch of highway over the muskeg faced due west if you travelled from Deschambault, Pelican Narrows, Jan Lake, Creighton or Hudson Bay. In the summer, the setting sun situated at road level blinded you in the eyes. In the winter, ice glare from the compacted snow on the road did likewise, but at 4 pm rather 8 pm. At times, it could be a dangerous drive in this locality.

Now back to the story. My Deschambault forestry assistant and I were at this 165 junction checking out a proposed logging block. We stopped for lunch, and the Deschambault forestry assistant recounted the experience of two young women who stood at that junction one

212

night as they hitchhiked to La Ronge. They were dropped off at this junction. This was a Friday evening.

A van turned off the main road at this 165 junction and stopped to pick them up. The young women climbed into the backseat and wedged themselves in since the van already was full of other Deschambault folk. Everyone got out in La Ronge and went their own way. Some months previously that van, and those people in it, were killed when the van was totaled on that dangerous stretch of Highway 6. They found this out after hitch hiking back to Deschambault and telling someone else of their good fortune. The two young women were totally shaken up by this experience. They probably remain that way today.

Mee-Toos

The Deschambault forestry activities became the basis for the formation of Mee-Toos, the new logging arm of PBCN, Prince Albert. 'Head office' on the PBCN urban reserve in Prince Albert administered this fledgling organization. Someone spent a large sum of money going to Silviba's chief competitor to get a new five-year plan made for Deschambault. The giggle was the competitor, with its digital mapping capacity, produced the same plan, but with a change of logo on its maps. Since it had no local knowledge of the forestry base, it was Silviba's plan anyway, but without an attached business plan.

The management of Mee-Toos (white guys) considered me as a threat to its ambitions, thus Silviba's mission at Deschambault was over. (In my opinion).

1992 Treaty Land Entitlement Framework Agreement

The treaty land entitlement agreement was signed in Saskatoon. I met Ovide Mercredi, who was then Grand Chief of the Assembly of First Nations. Prime Minister Mulroney was a signatory. I could not help but to feel that Mercredi would have done a better job as PM.

In 1993 Peter Ballantyne Cree Nation ratified the terms of the Agreement which allowed for an additional 234,248 acres of expansion to reserves and the formation of new ones. The government provided $62.4 million in settlement. This was paid out in full by the year 2003.

Silviba received a contract through the treaty commissioner's office to prepare a selection of lands. The primary choice of lands for Peter Ballantyne Cree Nation entitlement was the quality and quantity of timber available. The use of cost-benefit analyses confirmed the final selections. Our target lands were those supporting sawlog quality white spruces. The elders also recommended other selections that targeted spiritual and historic locations. We expanded some existing reserves, and proposed new reserves. I had to put values on Crown lands that, historically, had no market value. I used cost-benefit analysis to rank land parcels available. Sawtimber became the preferred value, together with secondary benefits. They included the opportunity for hunting and fishing lodges as part of the 'shoreland' values attached.

Peter Ballantyne Cree Nation Sandy Bay reserve

The sawmill here was equipped with a log carriage extension to saw longer logs. The homeland community of Sandy Bay bought it from the manufacturer in Oregon, USA, together with a Morbark electric-powered fencepost peeler. The community was set it up at Sandy Bay reserve (Peter Ballantyne Cree Nation) in 2002, or thereabouts. The project also included refurbishing a large bunkhouse. A business plan supported the funding application. This project had significant community economic impact. It could have employed 8 band members full-time. It collapsed before two weeks had elapsed. The saw carriage was damaged beyond repair the first day the crew graduated to operate the sawmill independently.

The peeler needed a 220-voltage power source, but this would have cost $7,000 to install. The project proposal had not included this cost. Why? I had planned for the diesel-powered machine. The band had substituted it to reduce maintenance. No money for power lines. Thus no posts were peeled. Heartbreak.

I had good accommodation while on site. I had a crew that cruised and mapped sawtimber, and fencepost sources. Thus, the raw wood supply was available. The funding also included purchase of a secondhand log skidder. The community bought lumber produced from the time the mill produced products. SaskPower's Island Falls hydroelectric dam close by would have bought significant lumber supplies.

Sandy Bay is a long way from mechanical support and spare parts. The economic development money was used up. In the future, we must contract for equipment maintenance/repair for any project proposal. Easier to say than done. There has to be money in the bank to pay for this.

Powerline clearing (2)

SaskPower issued contracts to both Pelican Narrows and Sandy Bay communities in 2003 for clearing a proposed powerline upgrade between the two communities. Silviba used the SaskPower drawings to develop a business plan. Henry Morin managed the project on-site for the Pelican Narrows stretch. Sandy Bay completed its stretch independently. SaskPower's usual practice with powerline clearing was to hire contractors with mowers attached to the back of a skidder. On this project there was too much standing timber. The Indians logged the material and sold it down south. In one sense, this was a graduation exercise. Henry Morin managed the project at the Pelican Narrows end.

Once more my job was liaison, and planning the work with Henry then planning the right-of- way units. Henry got the job done and paid for.

Ω

Chapter 19:

Silviba, 1987-88

1987

Silviba committed to potential forestry initiatives aimed at producing forestry jobs and businesses within Indian communities. Training in timber falling, machine operation, timber cruising and forest mapping, road location and cutblock layout. I did the business plans, training plans and sought out the suitable instructors. The communities took these forth for consideration by various funding agencies.

The Cree Nations' forestry crews remapped the forests on several reserves. Also, with these crews, Silviba undertook field surveys for Weyerhaeuser, including annual forest regeneration surveys. Indians had all the outdoor skills essential to staying on a field job until it completion; although they fell apart after 7 days away from their communities. The accuracy of their field measurements of all kinds was superior and held up well in Weyerhaeuser field audits of their work. On the takeover of the pulpmill and two sawmills, Weyerhaeuser agreed to a full inventory of all past clearcuts, even those that PAPCO had replanted. This increased the forest regeneration survey needs. The boys thrived in all weathers out in the forest.

One early spring, my crew of six from Pelican Narrows, set up camp to prepare for forest regeneration surveys. It covered a large block of Weyerhaeuser winter cuts just south of the Smoothstone River in Northern Saskatchewan. The roads, built over frost and compact snow, had just thawed out and became long stretches of sticky mud, puddles and muskeg. All-terrain vehicles were essential transport up there.

At 4 am, the world became alive with all sorts of singing, chattering, quacking, honking, and shrieking birdlife, all communicating at the top of their lungs – what a cacophony. The Sandhill cranes squawked their heads off as they selected mates and settled in to new nests. Oh yes, add in the frogs croaking in every pond, puddle and muskeg in the survey area, and beyond. I reminisce: "Have you ever tried to creep up to frogs croaking lustily as they try to attract a mate into their small patch of wetland? I never could. The croaking ceased the moment I even thought about it."

The crew learned to get up each morning just as the sun rose over the eastern horizon, cook breakfast, and then head out to the job. Once in the cutovers, higher and drier than the surrounding wetlands, the day's work began. Where the forest restocking was less than the required standard, Weyerhaeuser would send a tree planting crew in later. For some reason, the boys chased down and killed a Sandhill for supper one night. They burned off the feathers, but that was that. They preferred pork chops.

I was also working with Gene to develop proposals on behalf of Montreal Lake Development Corporation using Montreal Lake band members. Tree planting, tree spacing and logging became some of the projects undertaken on reserve. Gene also undertook a tree planting project on Montreal Lake band's other reserve, Little Red.

There were many projects continuing that early spring. The Montreal Lake chopstick study was completed. Never again will I allow others to place Silviba data under another letterhead. The errors created a problem for my professional reputation. Gene Kimbley had field projects underway at Montreal Lake. Silviba supervised tree planting projects on reserve at Chitek as well.

I note that on March 30, 1987 I sat in our workroom and mentally drifted. I was:

-Depressed.

-Frustrated

-Demoralised.

-Burned-out.

-Bored out of my skull.

-Anxious about the lack of cashflow.

Both Harry and Bob Romanchuk came through with $$$ spontaneously. It was working for small operators that kept me in Saskatchewan. I had prepared a brief for SCIFI (Saskatchewan Council of Independent Forest Industries). This laid out needs for an economically viable forestry mandate. Bob Romanchuk presented this to Eric Berntson and other cabinet ministers within the Devine government.

I found time to slip back to Lakehead University to run a forest fire simulator exercise in April.

Silviba did a major timber supply study for Newton Logging up the Hansen Lake Road (Hwy 106) south of Narrow Hills Provincial Park. East Trout, Elrick Lake, Stuart Lake were all included in the study. On May 13, 1987 the Elan forest fire exploded and burned over parts of Narrow Hills and through the South East Mossy Valley. This provoked an immediate change of plans. I sat at home watching the TV news when this fire blew up. The smoke column was visible from satellite photos as it travelled from southwest to northeast. It burned out much of the Narrow Hills Provincial Park in the process. As a result, Garry Newton gained timber salvage rights to some beautiful white spruce – bark scorched and dead – but there was nothing wrong with the lumber produced in the Newton sawmill at Cariboo Creek: 98% number 2 and better lumber grade. He employed eight men at the mill and contracted out the logging. Garry built the roads. Silviba did the cutting permit applications and flagged in the access roads. Fuel and groceries came up from Smeaton and Choiceland. I stayed in the bunkhouse at the sawmill. From there, I supervised logging and log scaling, as well as marking out cutblocks and access road locations. Garry had gone to hospital with appendicitis.

The Elan fire started as a lightning strike in a patch of muskeg. A helicopter on fire patrol discovered it. It had to fly back to base (hour 1 since detection), pick up a monsoon bucket, fly back to the fire (hour 2) and begin dropping water on to the fire from nearby lakes. By this time, sparks from the rapidly expanding fire spread the fire into the adjacent timber. The fire stopped a week later after burning through several thousand hectares of forest lands. May 13 was a date I had used in most of my fire simulator fire starts in boreal forests. And, as God, had scripted the arrival of a helicopter patrol with no initial attack capacity many times during my fire simulator years.

That spring, Silviba bid on large forest planning jobs for both the new OSB mill (oriented strand board) at Mafeking, Manitoba and for Weyerhaeuser in Prince Albert. However, larger consulting groups got both jobs.

My diary reflects the work underway. The busier I got, the more blank pages showed up. The balance of this year was filled up with:

- Indian work
- Operating plans for Newton Logging
- Amisk reserve
- Birch Portage bridge installation using the services of Stan Read
- Business development proposal for Provincial Forest Products Ltd in the Rural Municipality of Buckland
- Mentoring Gene Kimbley and Montreal Lake Cree Nation
- Private woodlot management planning
- The setup of that mysterious entity, Mirond Lake Forest Products Ltd sawmill in Birch Portage.

On November 2, a load of machinery from Morley Miller Equipment from Ontario arrived in Prince Albert at the C.O.D cost of $9,495.60. It was a power plant for the Mirond Lake Forest Products Ltd upgrade. Peter Ballantyne Cree Nation paid this. I arranged for transport to Deschambault. Various economic development funding arrangements contributed to the start up. Undoubtedly, Silviba did the funding proposals and business planning, also applications for training dollars.

This year was summed up by one word: 'mentoring'.

The diary does reference a visit to Calgary by Wendy Wiebe and I. She visited her folks and I met with Barry Manfield of Sylva Management Services. He based his company in Williams Lake, British Columbia. The discussions held probably lead to subcontract work for Silviba in British Columbia's Cariboo forest region later.

1988

That August job in BC

In 1988, I had to leave the province to find work. Coincidentally, the Canadian Institute of Forestry held its annual meeting in the Marlboro Hotel, less than a block from Silviba's office. I looked forward to this since I would meet many acquaintances and network with fellow foresters from across Canada. Unfortunately, just before this occasion, I was called to 100-Mile House in the BC Cariboo region to prepare the winter logging blocks for Weldwood, and missed the annual meeting. I had received my foresters ring from the Canadian Institute of Forestry in 1958 at UBC.

Before I left, I arranged to provide the coffee and donuts during the meeting sessions. I picked up Arthur Ballantyne, one of my crew, off the bus from Pelican Narrows, and drove to 100-Mile House in the old green suburban recovered from my Prince George days. This served as Silviba's first bush vehicle.

An item of note about that first journey back to BC. Arthur Ballantyne saw his first mountain on the approach to Jasper National Park. He stared up through the windshield all the way across the Park. He loved the mountains. At Mount Robson, I stopped and took a photo of Art standing in the parking lot at the base of the mountain. For once, the peak was in full sunlight. How many times had Mount Robson noted the presence of a Rocky Cree Indian at her feet over the last 10,000 years?

At the job site, located just east of the west boundary of Wells Gray Provincial Park, Art and I set up camp. To get there, we had to drive through an active logging show. The fallers had felled the Douglas fir trees and dropped them over a bank and on to the road below. Many of them smashed and broke. So, driving through these was difficult. I had to cleave to the external edge of the road that dropped vertically into a creek below. That was going in.

From the camp, we walked seven km through beautiful white spruce timber to reach the six cutblocks to timber cruise and map. It was a large job. Walking a fourteen km return trip through jungle each day slowed up progress. The timber was huge and scattered in places. On our return at the end of the project eight days later, we came across a logging crew fighting an escaped slash fire. We got out to help. Art picked up a Pulaski and piss-can,[46] and then went to work grubbing out the embers and hot spots at the edge of the timber. The supervisor came up to me and asked me where I could find more like Art. He was really disappointed when informed that both of us were on our way back to Saskatchewan. That was going out.

As we drove home through the Okanagan, Art was fascinated at the presence of fruit orchards, and considered whether or not to stay and pick fruit. We passed Mount Robson on the Yellowhead highway. Moonlight lit up the mountain peak. At 1 am the kiosk at the west end of Jasper National Park was closed. However, we drove through. I got us both home. Art stayed the night at my place north of Prince Albert before I dropped him off at the bus station the next day. Since the Forest Service checkers disputed my measurements, we had to return in late October and remeasure one block (the big timber).

The second item of note, on the second journey back to 100-Mile House, involved me. It was November, and as Art and I approached Jasper National Park going west, the clouds covered the mountains and the kiosk looked abandoned – no staff vehicle or other traffic – so I did not stop. But as I drove up the incline just before one of the junctions leading into Jasper, a Parks vehicle with all its lights flashing pulled me over. The ranger warned me I was now on the Park's database, and if I was ever caught again I would be in severe trouble. The Park ranger did not respond to my statement the kiosk was not staffed when I drove through.

Anyway, the second part of the job went much faster. We still had a return distance of fourteen km daily. We reconnoitred and flagged in the proposed main road and side spurs, and finished by flagging the cutblock boundaries. We discovered an old lava flow and a crater along the boundary between Weldwood's cutblocks and the west edge of Wells Gray Provincial Park. This fascinated both of us.

The job took another eight days. We flagged in the last landing on the last cutblock. As we began the seven km trek back to the vehicle for the last time a hurricane hit the valley along our path. It was both awesome and frightening. Huge trees crashed to the earth around us all the way back. Our safety was of deep concern. I had never experienced plow winds within deep forest.

We loaded up the van and trailer, and then started on the journey back down through the active logging. Weldwood had given us a portable radio. However, it was nighttime now and no one answered when I reported in that we were on the way out. The storm had blown trees over on to the road. All we had was an old blunt fire axe. We delimbed treetops at the road edge and worked our way along the roadway at the edge of a drop-off. At one point, Art had to lay a log along the edge of the drop off and guide me in the pitch black as I steered the front wheel over it while towing the trailer. We finally fought our way through the windfalls, then around the logging machine placed almost across the whole roadway, before reaching an unobstructed stretch. Weldwood did get on the radio to call us eventually. I told him that we were clear of the mess and on our way home.

The Harbour Dudgeon job

Silviba had one more job to do for Sylva Management that fall, probably between phases one and two of the Weldwood job. It was still late summer/early fall. The job entailed doing a timber cruise and logging layout in the Harbour Dudgeon watershed east of Vavenby. Access was up the North Adams River drainage. From there, we reached up into Harbour creek. The 'Small Business Enterprise Program' (SBEP) objective was to release logging areas and timber supplies for small logging operations. The BC Forest Service had redirected forest parcels from large area-based forest management licences. These remote pockets of poorer timber were hard to access.

Logging road construction was expensive because of the presence of steep terrain. This meant that benefits were fragile. It took money to do the entire preliminary field surveying, reports and environmental preharvest assessments. Habitat for birds, mammals and vegetation came up for discussion and documentation. Silviba and Sylva conducted the timber cruise, cutblock ribboning, road layout and PHSAs (preharvest silvicultural assessments). It was tough country to access. Steep roads with switchbacks would cost significant dollars to build. And the timber quality was not the best.

The BC Small Business Enterprise Program was 'a crock' in reality. Harbour Dudgeon was part of a large Weyerhaeuser forest management licence originally. Anyone bidding on that Harbour Dudgeon timber sale would have sold the timber to Weyerhaeuser anyway in return for financing dollars to build the roads. Also, there is limited small operator sawmill capacity in British Columbia, given the forest policies that protect the big guys – a familiar tune??

On the other hand, Weyerhaeuser in Saskatchewan did build the access roads needed to reach the timber. Small operators had free use of the roads. Weyerhaeuser had logged the lands accessed beforehand almost completely. This left small operators with dribs and drabs. Today it is all gone. No wonder Weyerhaeuser left Saskatchewan. So much for sustainable forest management.

Recently, Barry Manfield, owner of Sylva Management in Kamloops, told me the British Columbia government had turned Harbour Dudgeon into a provincial park. Probably, because no one bid on that timber sale anyway.

After Harbour Dudgeon, I returned home – just in time since snow was on the way. Again, it was 1 o'clock in the morning, and snowing, when I drove past the closed kiosk at the west edge of Jasper National Park. Before this, I had travelled up Highway 5 (North Thompson Highway) in the snow, arrived at the junction between Highway 5 and Highway 16 (Yellowhead Highway), then turned east on to Highway 16, through Jasper National Park before reaching Saskatchewan. Those who are familiar with this particular stretch of roadway will recall that one soon gets to a gate for use when the highway is closed because of impassable conditions. Once

224

you pass the gate, you continue uphill and around a corner. At this point, the roadway becomes two lanes in each direction, a slow lane and a passing lane. Stupidly, I passed a crawling semi loaded with pipe just before the roadway reduced to one lane on the descent beside the Terry Fox monument, with Terry Fox Mountain to the south. I met a semi coming up on the other side, travelling west and I barely squeezed between the two trucks – all the while battling visibility during a snowstorm – bloody stupid driving.

By the end of 1988 Silviba's cashflow was extinct. This holiday season looked bleak. However, we cashed in an RRSP. Silviba also received a Christmas Eve invoice payment by a small operator. That year, Marilyn Richmond did all of her Christmas shopping in Prince Albert's Golden Mall uptown. She started at 2 pm and finished at 4 pm, before all the stores closed for Christmas. Prayer performs miracles in tough times.

Canada-Saskatchewan 'Forest Resource Development Agreement (FRDA)

I was still working for the Saskatchewan government when I began the negotiations between Saskatchewan and Forestry Canada (later Canadian Forest Service). Several components were applicable to Saskatchewan:

- Private woodlot management contracts
- Forest research (Science)/growth and yield
- Forest research and field trials in silviculture and improved processes and results
- Forest ecosystem classification for Saskatchewan
- Disease and insect investigation and remediation
- Localising the Forest Fire Behaviour System for Saskatchewan
- Weather and climate impacts on Saskatchewan forest ecosystems
- Forest economics studies special to Saskatchewan

Over the years, Silviba took part in projects covering most of these components. The agreement was signed in 1985. For

225

Saskatchewan, studies into mixedwood forest management were critical, given that half of the commercial forest zone contained poplar.

Private landowners with forest properties received grants for presenting sustainable forest management plans. Silviba worked on properties from north of Meadow lake right through to Whitefox and Love, Saskatchewan. I used an Elan skidoo for overland travel in winter. It was an excellent work machine since it was so simple to fix when it broke down miles from anywhere. It was also a favorite of just about all Indian trappers.

These properties varied in forest productivity. East of Glaslyn, the tree crop was emaciated bluffs of aspen. It was difficult to develop enthusiasm for promoting a forest management plan. However, the owners presented all sorts of options, from replacing the aspen with a more valuable cover such as pine, to growing Saskatoon berries, to making firewood and furniture. These property owners were innovative in their thinking and had researched their options deeply. I completed inventories of all forest cover, added in access routes and trails, before writing up each individual plan. Once approved by each landowner, the plans were submitted to the Canadian Forest Service in Prince Albert for certification. There were tax implications for those whose plans included long-term sustainable management components.

There is something really precious about the atmosphere around a dining room table, or in the kitchen, listening to the farmstead owner explain his thoughts concerning high points of a projected forest management plan for his property. The wife would add her thoughts. More than once I was invited to stay for supper. I suppose that is partly why Saskatchewan remains in my blood.

The landowners with commercial forest crops close to Weyerhaeuser's sawmills had a ready market for their standing timber. Silviba mapped and cruised these private properties. The owners sold the logs to Weyerhaeuser. They replanted the logged ground with tree seedlings provided by the Canadian Forest Service. Later, sawmill companies, mainly Slocan Forest Products Ltd from British Columbia also purchased private logs from Saskatchewan.

Remember, our white spruce had exceptional wood qualities. A rail siding was built at Polwarth. At one time, 60 loaded log cars were parked in a siding at North Battleford waiting for the trip west.

It was not until I became involved in private land forestry that I understood the influence of past forest fires on the characteristics of new forests. The forest fires of 1917-22 erased the conifer gene pool from Pierceland in the west to Whitefox in the east. The only white spruce remaining stood scattered in valleys and around wet spots across the landscape[47]. These fires, fueled by land clearing and logging left behind a regenerated forest cover of aspen poplar. Why? Aspen poplar is rhizomous. Thus, the heat from the fires warmed the subsoil, and woke up the aspen rhizomes, which then sprouted. This is a common phenomenon in boreal Saskatchewan.

Not covered to any extent in the history books of Saskatchewan is that between 1904 and 1919, both Prince Albert Lumber Co. and the Ladder Lake Lumber Co. in Big River logged up to 100 million board feet annually from Dominion timberlands south of the Sturgeon Forest Reserve (now Prince Albert National Park) and east to Smeaton. This created enormous areas of logging slash and debris. Add in land clearing by settlers and the inevitable occurred. In fact, Prince Albert Lumber Company lost all its standing timber holdings in 1919 due to forest fires.[48] It later placed its sawmill on a barge and towed it to a location at The Pas in Manitoba. It metamorphosed into The Pas Lumber Co. and moved to Prince George in 1956 after the Saskatchewan government cancelled all of its timber permits on the east side of the province. The reason? The Saskatchewan Timber Board, a CCF government agency, needed the remaining wood supplies released for its own Crown timber operations.

The Forest Resource Development Agreement financed all forest management programming in Indian reserves. This provided Silviba with professional forester colleagues on the federal payroll who were available for support and advise, and who approved Silviba's invoices.

Ω

228

Chapter 20:

Silviba, 1989-90

Harbour Dudgeon continued

By January fourth, I was back at Clearwater, British Columbia, with Sylva Management to review the fieldwork completed previously. The Forest Service in Clearwater needed amendments and additions to the proposed logging boundaries and silviculture plans at Harbour Dudgeon. It asked for more Preharvest Silviculture[49] plots to complete the field coverage wanted. After meeting back in Kamloops to plan the next phase of the operation, I returned to Prince Albert.

PHSPs are part of any submission for logging rights, both in BC and Saskatchewan. Field plot sampling generates a total biophysical model:

- Soil profiles and geology.
- Inventory of all ground vegetation.
- Inventory and location of all rare plant species identified.
- Inventory of all shrub species.
- Decisions on habitat needs for all species of mammals and birds identified.
- Recording and mapping of all riparian exclusions.
- Analysis of all tree species present:
- Diameter at breast height (the standard tree size measurement).
- Bark thickness.

- Total age (using increment borers). Immature trees are excluded from logging.

The forest ecosystem is then zoned into ecotypes, digitally classified and mapped using GIS[50]. In turn, a postharvest plan is generated with the objective of renewing the forests logged. This becomes a need within the logging plans sent for government approval.

The Indian projects were advancing apace. These happened in stages - identity of a proposed project, whether it be logging, sawmilling, tree planting, road building or tree thinning. Chief and council chose each project through discussions and mentoring from me and Dale Reid.[51] Next, there were searches for capital funding, training dollars, and supervision. Two of the projects were on provincial Crown lands external to Indian reserves.

Silviba's office made up to 15 separate connections each day. This included liaison with up to six separate agencies, provincial and federal. These all wanted detailed plans, proposed cashflows, management arrangements, and proposed customers. They all wanted a postharvest forest renewal plan. Plans included two sawmills and the purchase of a poplar log home company at Weyakwin, all in Indian country.

The Indian world was difficult to keep 'on page' at times. Competition between reserves – that progressed to internal political manoeuvring – as well as inconsistent cashflow for Silviba, created stresses that almost knocked it off the map at times. I would put on my suit and tie, head down Central Avenue to the bank to brief the account representative on Silviba's bright outlook. Half of the City of Prince Albert knew exactly why I was dressed up. Given that cashflow was just as precarious on the Indian side, efforts by chief and council, and staff, to maintain Silviba's financial stability, critical on occasion, were essential. I regularly drove my Silviba truck up to 5,500 km monthly to cover all of our clients' needs.

Small operator support continued and working permits for logging and sawmill operations were applied for. Field surveys preceded this phase. The submissions included maps and volume summaries. Silviba relied on Indian forestry assistants for the fieldwork. I fired our remaining forestry technician because of absenteeism and drug use.

The need for non-Indian work meant that I travelled to contract jobs in other provinces, besides the Weyerhaeuser work locally, as well as a research project for Saskatchewan Research Council. This latter effort tested the capacity of remote satellite imagery to provide recognizable timber typing in a location within the commercial timber belt above the Saskatchewan Parklands. However, imagery in 1989 was too poor for acceptable forest interpretation. Another project involved the use of Lidar imagery in northern forests to map out the productive component for use with forest management. Unaccountably, this project was never completed.

Sandy Lake reserve forest operations plan, Northwest Ontario

In May, a firm of economic development consultants in Winnipeg contracted Silviba to prepare a forest operations plan on the Sandy Lake Indian reserve, north of Red Lake, Ontario. These were Ojibway Cree people, artistic, but saddled with a severe youth suicide rate. Access to the reserve is by air, or, in winter, by ice road – some 120 miles or so north of Sioux Lookout, the end of pavement. All freight arrives at Sandy Lake over the ice road:

- All building supplies and lumber sourced in Winnipeg
- All fuels
- All bulk groceries
- All school supplies, medical clinic, bulk material

The area was north of the Ontario forest inventory mapping zones and just east of the Manitoba border. Silviba did receive small-scale air photos, and used these on helicopter reconnaissance across the territory assigned by the Sandy Lake band. A man named Pardamus in the band office was my immediate connection with Chief and council. The economic development consultants were about to seek funding for the construction of a shopping centre at Sandy Lake. Silviba's mission was to set up a sawmill with logging and hauling capacity on the reserve. Note that sawmills start with the production of rough cut lumber: full dimensions. A rough cut 2 by 4 is exactly that prior to planing down and finishing (final size is 1.5" X 3.5"). Thus, the rough cut 2-by material is stronger than a planed 2-by something. It is unfortunate that Central Mortgage and Housing Corporation building codes apply to finished lumber sizes only. Remote Indian reserves in timber country should have the ability to saw their own construction lumber and build their own homes. This would save millions of dollars and create pride of ownership.

I returned to Prince Albert and completed the forest plan for Sandy Lake Indians. Then I returned to Sandy Lake once more via Winnipeg to present the forest operations plan to Chief and council. Pardamus put forward a few alterations then the firm of economic development consultants rounded up the funding to move in a portable sawmill and the associated logging equipment. See further accounts in 1990.

1989 July to December

The Enfor study[52]

Working for Silviba produced some stressful times. The Enfor study was one of these, a subcontract for all the field sampling needed. It involved field sampling of logged lands, stratified by ecotype, within Saskatchewan and Manitoba. We added a substratum, lands burned over. I recognise that folk who have read this far, may elect to skip this section. It is unavoidably complex, but is the basis for forest sampling. Foresters cannot measure every tree in the forest, so they select strata (groups) needing estimation within a chosen accuracy. For instance, you often hear of the results of a poll pointing, for example, to an average of 40 people out of a population of 1,000 people who are less than 5 feet tall, within +/- 15%, 19 times out of 20. We refer to this as the 95% confidence level. So, how many measurements to achieve this value +/ - 15%? It is all a matter of variance. The more sizes within that population, the more measurements (samples) to achieve the desired accuracy. If that population of 1,000 were all 5 feet tall, we would need just one sample height to decide the result. However, forestry values used in forest resource management are widely variable. This pertains to associated disciplines – mammals, birds, and fish as well. We use a 'rule of thumb' that says we need at least 30 plots prorated across a number of strata[53] to achieve an overall average of X cubic metres of standing timber volume with a sampling error of +/- 15%, 19 times out of 20, or at a confidence level of 95%. If you want to get a closer estimate than that, say +/- 7.5%, the number of samples needed expands exponentially, thus becomes uneconomic. So foresters live within the +/- 15% range just to save money.

The purpose of the study was to measure the quantity of residual biomass left from logging. This measure produced an estimate potential energy stored within this biomass, including humus. As usual, nothing was simple. The steps to take were:

- Determination of all companies with logged over lands, getting their cutover maps, then subdividing the areas by ecotype:
    - Pure poplar
    - Mixedwood poplar/spruce
    - Pure white spruce
    - Pure black spruce (with subhygric substratum)54
    - Jack pine types
    - Jack pine/black spruce ecotypes
- Contact with provincial forestry agencies to determine the timber volumes in cubic metres/ha available prior to timber harvesting
- Generation of a sampling plan: what percentage of the total area for each ecotype (stratum) was to be sampled
- Determination of the logging slash variables needed
- Each of these strata was then subdivided into:
    - a) handfalling/skidder logged or
    - b) mechanically logged and forwarded.
- Design of the sampling procedures, random or systematic plot distribution
- Calculation of all the costs involved, then setting up the work order details

The crew of eight from Peter Ballantyne and Montreal Lake reserves did the necessary measures in each plot. The minimum size of a piece of logging slash was 10 cm in width, to a length of 2 metres.[55] The job took 40 days, including office calculations. Each crew had a map depicting all plot locations. I remember Rod Morin, forest assistant from Pelican Narrows, losing a whole day stuck in the mud somewhere north of Turtleford.

The crew started work after budget approval that included transport, labour, food and accommodation expenses, all based on a predesignated number of plots. But, potential disaster struck eventually. The federal forester overseeing the project, his office was in Edmonton at the Northern Forest Research Centre, insisted the sampling accuracy was inadequate. This would have meant that that Silviba would have had to expand the number of plots, probably three times, for the same budget. Silviba would have gone broke before it was out of the gate. Fortunately, Silviba's sampling plan remained intact. Both the contractor's rep, a forester with a PhD, and a federal forest scientist associate, managed to persuade the federal forester the sampling plan fit the specifications desired. However, this was a 'near miss'. Even foresters become confused by statistics.

A second part of the Enfor study was an ecological analysis performed by Hamish Kimmins PhD, the renowned forest ecology expert from the University of British Columbia. He showed that no ecological damage would result from this harvesting of the residual biomass. That was one criticism of the study at the start.

The results suggested that logging slash in Saskatchewan and Manitoba stored enough energy to simulate the explosion of several hydrogen bombs.

1989 fall
The fall was just as busy. Some project funds were received and equipment purchased. Some training plans were approved, other proposals rewritten four times. Among the funding agencies in play:

- S.I.E.F = Saskatchewan Indian Equity Fund
- CJS = Canadian Job Strategy
- CEIC = Canadian Employment Insurance Commission
- NEDSA = Northern Economic Development Agency
- ABDF = Aboriginal Business Development Fund

- CAEDS = Canadian Aboriginal Development Strategy (created in 1989)
- SaskPower Northern Enterprise Fund (bought a log skidder for Peter Ballantyne Cree Nation)
- NIAC = Northern Indian Agriculture Committee
- Northern Enterprise Fund
- Visions North
- Canadian Forest Service within FRDA
- Indian and Northern Affairs Canada/Resource Access Negotiations (RAN)

It would be more sensible to set up 'one stop shopping'.

All the foregoing funds were activated with varying degrees of success to accomplish Peter Ballantyne Cree Nation forestry goals. Dale Reid pointed me in the right direction. We did the research and paperwork.

Keyano

Silviba spent much of its energy consulting on forestry opportunities with the Beaver River Community Futures agency at Buffalo Narrows in northwest Saskatchewan, north of Meadow Lake. This was Métis country, remote, broken terrain, poor road access, and beautiful lakes. But, it had a long history of development going back 250 years, starting out with jobs in the fur trade including transporting goods on the Churchill River and commercial fishing on adjacent lakes. French-Canadian traders left their mark by establishing communities such as Beauval and Ile a la Crosse (http://www.sakitawak.ca/)

Consideration of logging opportunities, sawmills, fenceposts, peat harvesting and electric cogeneration became part of business proposals. The respective funding agencies became involved. It was a long-drawn-out exercise; however, I made many friends and connections. This was a follow up to my first 1984 mission promoting indigenous involvement in managing the forest lands in the Northwest, but on the Métis side of the equation. My most notable memory is of Louis Morin of Jans Bay on Canoe Lake. He was known affectionately as 'Mr. Mechanical'. He had lost a hand that he replaced with a mechanical device. This never held him up for one moment. More of him later. I sat in his kitchen for lunch one day. Precious.

The management expertise behind Beaver River Community Futures was a retired bank manager in Meadow Lake, as well as his sidekick, a retired car dealer from Saskatoon. The latter was seeking a fitting opportunity for investment into a likely business prospect. The anchor project was the proposed electric cogeneration plant at Buffalo Narrows.[56] SaskPower had contracted for the supply of 15 megawatts at a stated price. I had fortuitously found a designer and supplier of the equipment needed, then produced a raw wood supply plan with all harvesting, transport and related costs. The project was doable except that SaskPower fiddled with the proposed price for delivered power. This meant the forecast cashflows changed several times. Finally, the financial agencies and banks behind this project withdrew. Perhaps SaskPower achieved its end. No intention of ever going through with this first time ever contract for independent power. Turf protection once more? SaskPower's Tony Harras? Or was the powerline capacity leading to the electrical grid too expensive to upgrade from Buffalo Narrows. In reality, the project was fifteen years ahead of its time.

The failure of this project resulted in the failure of three other major proposals that would have revitalized the northwest communities. Eventually an experienced carpetbagger, ex Kitsaki Development Corporation manager in La Ronge moved in and set up his kingdom. His genius was in finding capital dollars for projects. Over time, he drained the coffers of the Western Economic Diversification Program regularly. He joined with less senior carpetbaggers to make a nest that lasted for years – and never carried out one forestry project for the northwest side. He was American. I was both disgusted with me because of my lack of expertise in finding money, and also with Honourable Ralph Goodale, the regional federal minister in charge, who doled it out at one million dollars per request for insiders. It felt as if Fortunato Pacios-Rivera had returned to Earth in disguise.

Eventually, the northwest trappers insisted on a charge of 50 cents/cubic metre for all timber logged along a trapline as compensation.

Provincial Forest Products Ltd, Rural Municipality of Buckland

Provincial Forest Products Ltd was located at Spruce Home in the RM of Buckland, north of Prince Albert on Highway Two North. Barry DeVocht, the company owner, and brilliant entrepreneur, operated within restricted Forestry Branch timber quotas. Yet, his business persistence continuously evolved. He had Silviba do up a business plan for presentation to the Federal Business Development Bank in Saskatoon, but once more bureaucracy intervened. No extra wood, no business loans, thus no upgrade to the existing sawmill. One common sense upgrade would have been to install a remanufacturing line. This would have increased value to the company's product array, improved cashflow and added more jobs. It was not to be. The Saskatchewan Forestry Branch denied the additional trees required, thus the Federal Business Development Bank declined to approve the business plan. All so shortsighted and economically destructive. PFP's first sawmill produced much work for Silviba over the next few years – an incessant mentoring exercise.

That summer, Silviba conducted an inventory of all discarded power poles throughout central Saskatchewan for PFP. These were the result of SaskPower's RUD program (Rural Underground Distribution). Barry ran test loads through his sawmill, but the lumber produced was too weatherworn and with too much embedded iron.

Later, PFP was the first small operator to saw poplar commercially. Barry sawed the material into 4X4 bolts for supporting pipeline pipe before it rested on the ground. He sent many truckloads of poplar squares into Alberta.

Barry DeVocht was constantly scheming to survive in the face of roadblocks put up directly by Forestry Branch, and indirectly through Weyerhaeuser. Somehow though, his slab pile beside his sawmill grew higher and higher over time. Eventually, the municipality, egged on by the neighbours, threatened to take him to court for contravening a bylaw. But by this time the stored energy sitting in that slab pile threatened his sawmill, shop and office. One dark night, in a drizzle, he lit it all up and luckily was left with just a mess the next morning. All structures were untouched.

In Conclusion

All this while, I was teaching myself how to use Smart 2 software on Silviba's new computer, with a dot matrix printer attached. I had financed the purchase from RadioShack with a Bencharge loan at 32% interest. Says something for our banking system. Microsoft Office was not to show up for some years. In September, Silviba looked into forestry opportunities for Peter Ballantyne Cree Nation on its Southend reserve, chiefly, a wood supply to feed a sawmill. In October, Weyerhaeuser asked Silviba to conduct forest regeneration surveys on a number of key areas. Weyerhaeuser needed the information gathered to complete its tree planting estimates for the following spring. As well, I made two trips back to Sandy Lake Indian reserve in Ontario, and investigated forestry opportunities at Sturgis, SK. In November, I joined Peter Ballantyne representatives in Vancouver to attend the annual National Aboriginal Forestry Association symposium. My specific assignment was to man the Peter Ballantyne booth, with a montage of Mirond Lake Forest Products displayed as a backdrop.

1990 January to June

In the first few months, Silviba was deep into fieldwork as it checked out and surveyed timber supply for Provincial Forest Products Ltd. It also mentored Peter Ballantyne Cree Nation during the development of its forestry opportunities. Training proposals, then to Petawawa National Forest Institute northwest of Ottawa to attend seminars on forest fire research, and a tour of forest silviculture trials going back to 1918, all took up time between mentoring episodes. I visited PAMAP in Victoria in January to scope out this forestry GIS software product for use with Silviba projects.

In March, I attended a Geographic Information Systems (GIS) introductory course in Vancouver, followed up with a week in Regina at a seminar for GIS aspirants, at the University of Regina put on by Dr. Dave Sauchyn. Meanwhile, ongoing support of small operators and Mirond Lake Forest Products Ltd in Deschambault continued unabated. This latter entity demonstrated a more complex forest development opportunity for Peter Ballantyne Cree Nation. In April, a grand opening of the new sawmill in Deschambault attracted dignitaries from both the Indian and White worlds. This mill was designed to saw up to 15,000 board feet per day of rough lumber. This is a relatively modest output in terms of sawmill capacity, but due to management constraints and absenteeism, there were times that it was shut down for months.

This sawmill project was reaching into a higher level of influence within the band. The director of economic development for PBCN, Ron Ray, a resident of Sandy Bay, was now involved through more sophisticated funding agendas. He based his office on Peter Ballantyne's urban reserve in Prince Albert, rather than back at the band's traditional home at Pelican Narrows. He gave me the job of putting a five-year business plan together for Mirond Lake Forest Products Ltd as a precursor to a NEF funding application (Northern Economic Fund). Dale Reid, the chief's right hand man, initiated economic development within homeland communities. Here, he had more control over cashflow, and could protect his personal ability to survive. Dale dealt with the internal world of reserve level politics, while Ron Ray dealt globally: two opposing constituencies. Eventually, Silviba found itself in the middle.

Ω

In April, Silviba undertook a large forest regeneration survey on Weyerhaeuser's cutovers using its Indian forestry assistants. This went on until the end of May. However, the office compilations and data entries lasted until the end of June. I used Clarence Michel (Nick Nick) for some of the office compilation. All these young people lived up to 450 km away from the office. Once the fieldwork finished they had no wheels, and were unavailable. This was unfortunate since Clarence learned so fast. He did use the bus service. Even so, he was an unreliable presence at the office. I did loan his upcountry crew a vehicle to get home occasionally, but it was not a good business practice to have a vehicle unattended on the reserve.

1990, July to December.
The next few months involved hectic moments assembling various funding applications:

- The continuing Ron Ray request for preparing a NEF funding application

239

- Bert Luckhurst's request to put together a cost-revenue projection over five years for the Sandy Lake, Ontario project

This meant the forecasting of total revenues and cost of goods sold. In turn, this of projected orders from committed buyers of lumber products.

- A plan for management and administration
- A clear direction of ownership
- A market analysis and confirmation of cost needs for machinery and equipment
- A confirmed commitment to share the risk with other funding agencies

This latter need was the most time-consuming and laborious. Not to forget ongoing projects with Newton Logging of Cariboo Creek, north of Smeaton, a new request for a five-year timber supply plan from Harry Romanchuk of R&T Trucking. Harry ran a sawmill on the White Star Road in the Rural Municipality of Buckland.

All of this forestry business planning used various cost sources through work done for small operators. Jackie Ross kept me in touch with delivered log prices, Barry DeVocht briefed me on lumber prices *fob*[57], as did Garry Newton. Add in the funding application for Sandy Lake reserve through AEDP (Aboriginal Economic Development Program).

- Nikotawsik Forest Products Ltd application through SEDCO (Saskatchewan Economic Development Corporation) for funding
- Continuing efforts by Provincial Forest Products Ltd to get long-term financing

All of these efforts needed five-year business plans. These, in turn, triggered the need for an assured five-year timber supply. This involved provincial authorities.

Each timber supply plan incorporated field sampling (timber cruising), mapping and data compilations in the office. And then, there were farm woodlot plans at any time. Include that with a large right-of-way clearing project for a new transmission line from Pelican Narrows through to the Two Rivers junction to the west awarded to Peter Ballantyne Cree Nation (I did the project costing), as well as a powerline right- of-way clearing project at Deschambault. Silviba's workload approached saturation.

This was interspersed with jobs for both Shoal Lake and Red Earth Indian bands (technically Swampy Cree). We started a forest management plan for the Big River band. The year ended with the need for more office space, a new banker, attendance at a ten-day GIS training course in Victoria for me, Wendy Wiebe and Cathy Maruschak, funded by Canada Employment and Insurance Commission. While in Victoria, Wendy and I interviewed a new UBC forestry graduate. This was all premised on a five-year business plan put together by me for Silviba Services Ltd.. Oh yes, Silviba finalized the purchase of the PAMAP GIS software program.

Ω

Chapter 21:

Silviba, 1991-92

In 1990, at an international environmental conference in Rio de Janeiro, Brazil, the delegates lambasted Canada for its non-sustainable forest management practices. In the aftermath, the Canadian government of the day called a meeting in Winnipeg. The Prince Albert district manager for the Canadian Forest Service was one of an array of public forestry officials from across the nation who took part in a debate on how to combat the image created at Rio. They determined that Canada should create a model forest program by inviting bids from all interested provinces. The Canadian Forest Service provided application guidelines. The participating provinces went home to study the costs and benefits of setting up a model forest on their turf, subject to the receipt of acceptable bids. The federal budget for this covered ten potential provincial model forests.

I remember the first meeting with all interested parties at Prince Albert in a ground floor meeting room at the Marlboro hotel. Vic Begrand, the Canadian Forest Service manager for Saskatchewan laid out the contest rules. Should this be a 'go' or 'no go'? The various participants from Weyerhaeuser (the big gun), to the provincial forestry department (a confused gun), to the Canadian Forest Service (major bureaucratic gun) and members of the Montreal Lake Cree Nation agreed to bid on a model forest program for Prince Albert. This was in May, 1991.

Promptly, Jack Spencer, a forester with Weyerhaeuser Saskatchewan assumed the task of writing up a draft agreement. For the Indian side, I served as mentor and coach for Montreal Lake Cree

243

Nation through Gene Kimbley, its human resource for forestry. Noland Henderson, MLCN Chief then, contributed the proposed eagle logo and Prince Albert Model Forest (PAMF) mission statement (short form) in Cree. I then worked with Jack to blend the Indian wishes into the final document draft. Although the final PAMF agreement included signatories from both La Ronge Indian Band and Prince Albert Grand Council, their involvement faded in time. By early 1992, all parties had signed up. At one point, I chased down Chief Harry Cook in La Ronge to get his signature. Other signatories included Saskatchewan Research Council and Prince Albert National Park.

Before proceeding, one is reminded that the management of Canada's forests lies with provincial jurisdictions. First Nations traditional lands in Saskatchewan disappeared because of the Natural Resources Transfer Act of 1930. Thus, the province of Saskatchewan manages the forests in the province. The impact of resource lands transfer from Federal to provincial responsibility on the Indian nations was profound. They lost their traditional landbase external to the Indian reserves. Access to natural wealth was refuted, even though Indian treaties signed prior to 1930 did try to address domestic issues. The language was archaic, and the intent too vague. The treaties never included representation from provincial governments because these were yet to be born. Thus, Saskatchewan Forestry Branch was less than enthusiastic to share its jurisdiction with Prince Albert Model Forest. The provincial bureaucracy could not bully partners such as Prince Albert National Park or Weyerhaeuser, or even Indians; thus, it took out its *angst* on poor old Silviba Services Ltd.

Ω

In February, Montreal Lake Chief Noland Henderson had me go with him to Ottawa to attend the National Indian Forestry Congress annual meeting. I wrote Noland's speech and contributed to various working groups.

Ω

Silviba was busy with Weyerhaeuser forest regeneration surveys and small operator cutplans. I wrote the proposal to produce a forest fire suppression Resource Access Negotiations funding application for Prince Albert Grand Council. This was the result of my mentoring efforts with Gene Kimbley, and through him, Montreal Lake Cree Nation. The search for a program manager with connections to Saskatchewan Forest Fire Management branch became a priority. Wages were a top up of welfare payments. A potential employee needed to be on welfare at the time of hiring. Since this was a federal (INAC) responsibility, Saskatchewan got its Indian fire crews at a subsidized rate. One observation, after the program got underway, was that many good young Indian folk, men and women, were denied employment. I wonder if any of this changed.

On June 15, Suzanne Magnin started as the GIS operator at Silviba.

Ω

In November, Silviba had two more assignments: First, measuring trespassed log volumes from a farmer's white spruce bluff on his farm; Second, a logging waste assessment after logging poplar on the Mistik Forest Management Agreement lands at Meadow Lake. This latter job needed the design of a sophisticated waste wood sampling plan before collecting data. It took a week to sample the areas chosen and measure the data needed, followed up with a week in the office to calculate, then summarise the results. This became the basis for Mistik's research into developing a heat source for the generation of electricity. I thrived on a project such as this. The challenge of the job released zeal and enthusiasm, and Mistik paid invoices promptly.

This was followed up in December with another project for Mistik Management. Since Mistik was the wood supplier to the Meadow Lake poplar pulpmill, Silviba orchestrated the salvage of a vast area of infected poplar within Greenwater Provincial Park. The previous summer forest tent caterpillar severely devastated the Greenwater Park trees. The Park was devastated; poplar trees were sick, and Mistik needed the wood. My crew discovered a patch of the original shortgrass prairie during mapping forest damage. This prompted some research into the movement of forest boundaries over time. Two hundred years ago, the climate was dry. Trees could not survive, so grasslands held on. Then came a mini ice age and humid summers. The treeline moved south and drowned out this shortgrass prairie ecosystem. The current patch is in danger of reinvasion. Fire was the only tool if this little patch of shortgrass prairie was to stay *in situ.*

Ω

Back in the summer, I travelled up Reindeer Lake with Henry Morin. On the way up, we fished for lake trout. We stopped for a wonderful feast of bannock at the Clark family summer camp. I ribboned out the boundaries for a new Peter Ballantyne reserve at Co-op Point (Kinoosao), before returning to Southend by boat the next day in a howling storm. Henry's dad, Phillip Morin, knew what to do as he stayed in the lee of available islands on the way home.

I cannot continue without referring back to the Clark family and Reindeer Lake. Mrs. Clark is Cree from the tip of her toenails to the top of her head. Yet her skin colour is fair. Folks at Co-op Point told me her coloration is passed down from a white explorer, perhaps from Iceland, more than a hundred years ago. Reindeer Lake is awesome. From storms creating whitecap waves to a summer day of absolute calm – the ninth largest lake in Canada – the experience is noteworthy. On that trip up to Co-op Point, the water was glassy. The islands ahead shimmered in the heat. It was surreal. We could not tell distances.

Ω

Chapter 22:

Silviba, 1993

January to December

What a dog's breakfast: In January and February, Prince Albert Tribal Council searched for a forest fire coordinator to run the forest-fire suppression program to provide forest fire suppression crews to Saskatchewan Forest Fire Management Branch.

-On February 1, Todd Bell, Silviba's forester, gave his notice for May 1.

-Gene Kimbley pressured me steadily to draft up his coordinator position with the Prince Albert Model Forest.

-Peter Ballantyne Indians and I reconnoitred the backcountry to discover the suitability of certain locations as reserve parcels.

-Mistik Management in Meadow Lake provoked a switch of a portion of the forest land base from the Weyerhaeuser Forest Management Agreement over to Mistik's FMA. Silviba produced a forest inventory summary for the lands in question, then displayed this on a GIS map.

-Jack Custer asked that I find him a 200-amp breaker for the sawmill at Deschambault as well as find him a trainer for his newly received sawmill edger (thanks to Stan Read's connections).

-More stuff between PBCN, the government and Jack Falk over the Indian IRMA proposal; including a draft memorandum of understanding between the band, Saskatchewan government and Jack Falk.

And then there was a tour with lumber brokers for small operator sawmills. Followed by:

-Silviba's bid on a Prince Albert National Park (PANP) project to survey balsam woolly aphid infestations. More mentoring of all involved parties connected to the Prince Albert Model Forest. I wrote the chief's speech for the occasion of the grand opening of the Prince Albert Model Forest, including introducing the new executive director, Dr. Thomas Bouman.

-Jackie Ross and Barry DeVocht (Provincial Forest Products Ltd) made a deal with Jack Custer at Deschambault to log reserve timber for shipment to Barry's sawmill in the Rural Municipality of Buckland. It was Silviba's job to process all the paperwork, lay out the logging blocks complete with access roads, and gain the cutting approvals from two agencies: Indian and Northern Affairs Canada in Prince Albert, and the Canadian Forest Service, also in Prince Albert. Remember, Indian reserve lands are federal jurisdiction.

-Fresh energy by The Métis from the Saskatchewan Northwest communities needed planning support from Silviba to explore a sawmill site – with electric power generation capacity using wastewood (fire-killed timber). This would also include an agreement from Mistik to release wastewood.

-At the end of April, Weyerhaeuser told Silviba that Weyerhaeuser Timberlands would switch from PAMAP to Terrasoft. This was a competing GIS software product, built by former Macmillan and Bloedel foresters in Nanaimo, BC. This was disastrous news since it meant that Silviba could no longer communicate with Weyerhaeuser's GIS, thus no more contracts. Devastating.

Ω

On June 1, Silviba was despondent over the signs of a company failure. The GIS news, combined with an empty bank account, made for an 'iffy' start to June. However, the Company was still functioning when I took Peter Ballantyne Chief Ron Michel and Council, with staff, on a sawmill tour of certain western Canadian sawmills, starting June 12. They travelled on a Prince Albert Northern Buslines tour bus. They started at the Sundance mill in Edson, then went through to Merritt, BC to view two more sawmills, before travelling through to Vancouver to meet some high-priced economic development types from Indian Affairs downtown. The topic was the PBCN Resource Access Negotiation to cover the cost of current negotiations for the proposed Integrated Resource Management Agreement with the province.

Dale Reid, the chief's executive assistant orchestrated the whole trip. He used me for the technical set up. We travelled separately in a Silviba pickup. Two items of record remained with me:

The bus stopped at Hope, BC on the way home for a washroom break and a restaurant snack. Everyone had pie and ice cream. The Cree tongue, however, had significant trouble with pronouncing 'bumbleberry'. After all the giggles subsided, they ordered what sounded like 'burb-burbelly' pie.

We stopped that first night on the return trip in the Roger's Pass at a hotel. Early the next morning, I took the reluctant 'keeners' for a hike behind the chalet, and up into an alpine valley. It was a beautiful morning; they almost got up to the ridge top before returning to the bus. Years later, that same valley was the site of an avalanche that killed many schoolchildren from Calgary.

Air Weapons Range fire salvage

"Dry lightning storms in mid-May triggered multi-fire starts in northern Alberta and Saskatchewan. These required mobilizing more land-based air tankers into Saskatchewan. On June 3, Saskatchewan reported a fire originating on the Primrose Air Weapons Range (Deer Fire) at 47,000 ha. This fire persisted for most of the 1993 season, with its final size reported more than 300,000 ha." (University of Freiberg – International Forest Fire Research group).

The fire started from NATO aircraft on their annual training excursion from the RCAF airbase at Cold Lake, Alberta. They used phosphorus bombs that ignited forest fires and burned over vast stands of commercial sawlog and fencepost timber – none of it available to small operators, or large ones, for that matter. The fire started on the south side of Primrose Lake, and then ran in a northeast direction pushed by a southwest wind[58]. The Air Weapons Range, also known better as the Primrose Air Weapons Range, is Saskatchewan provincial forestlands, but leased to the Canadian Defence department for training. Through the lease agreement, no one gets access to the Primrose range, not even forest firefighters. The U

S Air Force had permission each spring for missile launching from Arctic airspace onto targets within the weapons range using 'cruise' missiles carried by B-52s. Access to the Deer fire from outside the range was *verboten*. All access questions went to the CFB air base commander at Cold Lake.

And, so started six months of intrigue, stonewalling, lies, turf warfare, First Nations recalcitrance and Métis ambivalence. On June 22, Barry DeVocht, Jackie Ross and I met to explore the possibility of Air Weapons Range timber salvage. Remember, Silviba Services needed reliable cashflow. Barry DeVocht, Provincial Forest Products Ltd (PFP) needed more timber. Jackie Ross (JR), a treaty Indian from Montreal Lake Cree Nation – he ran a logging company – needed work for the following winter.

ProForest Ltd (1)

The first item on the agenda, an estimate of the volume of timber available by species, became Silviba's responsibility. During the analyses needed, the forest inventory section of Saskatchewan Environment & Resource Management cooperated diligently. Barry DeVocht, Jackie Ross and I, and the lawyer, Peter V. Abrametz, whose office was just down the block from Silviba, as well as a moneyman, Eric Brown, economic development consultant from Saskatoon, formed a company: ProForest Ltd. I had my lawyer create a new company, Glim Holdings Inc. to hold my ProForest shares. These would have to originate with professional services instead of cash.

On July 12, Saskatchewan Environment & Resource Management enabled ProForest to log salvage timber off the Air Weapons Range. This was of critical help since no business plan could advance without a guarantee of timber supply.

Carrier Lumber Ltd (1)

Carrier Lumber, based in Prince George, had an exceptional advantage in that it had the capacity to take its portable sawmills to the forest, rather than truck the logs a long distance. This meant the truck haul had an economic advantage – hauling out finished lumber, not lower value logs. Later, Carrier Lumber (Kordyban family from Prince George) comes to life in a big way in Saskatchewan. As well as the portable mills that it designed and built itself, Carrier also ran a used equipment business. Thus, it could also send logging equipment out with its portable mills, especially large D8 and D9 caterpillar tractors, bunkhouses and cookhouses.

Barry DeVocht had his fingers in several pots. He told me about Carrier's high-grade portable sawmills used for timber salvage. Carrier had started out as a timber salvage company salvaging all the timber to be drowned on the proposed Williston Lake in British Columbia, north of Prince George. He made the phone call to Bill Kordyban Senior's office in Prince George. Barry gave Bill a summary of the fire on the Air Weapons Range.

On July 28, the Kordybans flew into Edmonton in their Cessna Caravan aircraft (high wing, single engine, with a passenger capacity of 10-14 passengers). Barry DeVocht, Jackie Ross and I joined them to fly over the Primrose Air Weapons Range. Beforehand, Murray Little, the director of forestry in Saskatchewan Environment & Resource Management sought permission on our behalf to fly over the fire area. We had to squeeze in our survey between NATO training sorties. What was the character of the ground access? Was the timber salvageable? Was it of sufficient quality (size) and volume? Where would the Kordybans set up the mill? We only had ½ hour to get our survey completed. I had brought along a Saskatchewan forest inventory map. With this, we designed a flying pattern covering the potential timber salvage locations. We passed over into the Air Weapons Range at 4 pm precisely.

The Kordybans met later with ProForest on October 11. We all travelled up To Keeley Lake lodge for a 'heart-to-heart' meeting. I had chosen a sawmill site on the Air Weapons Range from the specifications provided by Carrier. The next day, they flew the Air Weapons Range once more, this time with Floyd Wilson, the regional forester for Saskatchewan Environment & Resource Management, Meadow Lake, aboard. The previous inspection in July was a hurried trip because of a narrow airtime window. It will be shown that Floyd was part of the Mistik scheme to keep ProForest off the Air Weapons Range. I presented the estimates of salvage volume based on the Saskatchewan forest inventory.

ProForest (2)

A meeting on August 25 between ProForest and the board of NorSask Forest Products Ltd occurred in the sawmill boardroom in Meadow Lake. Its purpose was twofold. First, to determine ProForest capacity to deliver logs to the NorSask mill in Meadow Lake. Second, to discover the extent of the ProForest game plan and its potential for investment. NorSask was owned by the sawmill employees and the Meadow Lake Tribal Council. Alan Brander was a new sawmilll manager at the time.

A meeting between ProForest and Mistik Management (managers of the Forest Management Agreement area blocking access to the Air Weapons Range), occurred on September 9 in Meadow Lake. Barry Peel, the general manager of Mistik, chaired the meeting. This was a significant hurdle meeting and the atmosphere was toxic. Although ProForest had government permission to log on the Air Weapons Range, Mistik considered this to be its turf. Because of its connections on Saskatchewan's west side, it could block ProForest capacity to continue with its salvage plans. This would have affected ProForest capacity to raise investment funds – what a switch. Up to now it was government that blocked access to timber supplies.

A meeting took place on September 22 in Saskatoon between ProForest and Chief Guy Lariviere of Canoe Lake Cree Nation, and Leonard Iron, a councilor. Mike Blackmon, representative of Beaver River Community Futures (the Métis connection) also joined the meeting. The Deer fire had burned right into territory viewed by the Canoe Lake Cree Nation on the northeast side of the fire as being their turf. They considered this their traditional land base. However, in a counterclaim, the Métis communities of Jans Bay and Buffalo Narrows were equally insistent the fire had burned into their traditional hunting and trapping blocks. This was the situation facing ProForest as it began the job of organising forest fire salvage up on the weapons range. Anyway, the elders of this Cree nation wanted no part of a clearcut logging scenario – another hurdle.

Earlier, during the birth of Mistik Management based in Meadow Lake, the northern communities had blocked logging roads.

Their concern centred on clearcut logging. David Suzuki was one prominent ingredient of this roadblock.

On October 14, Jim Engdahl, the moneyman in Eric Brown's economic development consultancy in Saskatoon, warned ProForest that both Mistik and NorSask were scheming to keep it out of the Air Weapons Range.

On October 25, ProForest began to strategise a plan to bypass Mistik.

On December 12, Silviba filed the Carrier sawmill site plan with Saskatchewan Environment & Resource Management, subject to approval by Canadian Forces Base, Cold Lake.

Twice in December, Mistik told ProForest there would be no Air Weapons Range salvage timber available. Ironically, there would have been no Air Weapons Range salvage project if ProForest had not stepped up to the plate. Forestry Branch had no plan of encouraging a salvage option beforehand. Mistik and NorSask were unenthusiastic at the thought of rearranging all their wood supply planning. Also, fire salvage hastened the risk of char in the wood chips. The only cure for this was to double-debark all timber entering the NorSask sawmill. NorSask would lose its chip supply contract with Weyerhaeuser if they found char in the digesters. Forestry Branch should have put its foot down: no more green timber until Weapons Range timber was salvaged. However, it tiptoed around the issue in a visible absence of leadership.

On December 23, Jim Engdahl submitted Pro Forest's business plan, as developed by Eric Brown and I, to Gary Benson of The Saskatchewan Government Growth Fund (SGGF). This entity was a subsidiary of CIC (Crown Investments Corporation) based in Regina.

Shortly after, ProForest met with the elders at Canoe Lake First Nation for a continuing discussion on clearcutting and the use of mechanical harvesting. ProForest needed their approval before moving its logging equipment into the bush. Time was of the essence. Since it was now past freeze up; access roads and muskeg access needed freezing in. Mistik loggers were already on site with no hurdles to jump over.

And all retired for a lousy Christmas.

Ω

Chapter 23:

Silviba, 1994

ProForest (3)

Everyone connected to ProForest, teetered on the edge of hysteria as it tried to find investment funds. It finally received permission from both the Métis and Canoe Lake Indians, Mistik and the base commander at Cold Lake, to log on the Air Weapons Range. Thus, ProForest could set up a logging camp, and sawmill site. In January, the Saskatchewan Government Growth Fund wanted a due diligence on the supply of available salvage timber. This cost $5,000 as well as the helicopter cost amounting to $4,000. Of course, the report issued dumped on Silviba's estimates without understanding that Silviba's timber supply estimate for Air Weapons Range salvage incorporated some smaller trees the consultant left out. I designed the data sampling to suit the log profile of a 'hewsaw'.[59]

An intensive search for loggers and logging equipment, as well as a cookhouse and bunkhouses was well underway by mid-January, even though the permissions had yet to arrive. One underlines the winter logging season in muskeg country has a narrow window: January 1 to March 1. The roads freeze in before Christmas, and the loggers and truckers follow after New Year. Then they work ten days on and four off, in two shifts of 10 hours. The mechanics maintain and repair the equipment during that spare four-hour period. One has to plan for at least one cold period towards the end of January. Once the daily temperature falls below -30 deg. Celsius, operators park their machines due to the chance of broken parts from brittle metal and frozen hydraulics. Even so, most bush outfits continue with some aspect of the logging. If all the decked logs at the landings are not loaded out by March 5, they stayed there, not paid for, until the following winter. At times, nature may extend the season for a while, but never count on this.

Freezing in a logging road that crosses muskeg starts in mid-November when snowmobiles traverse the road route to drive down the frost layer. After that, a log skidder will traverse that same route, and drive the frost line down to a greater depth. Finally, A D7 wide pad cat will drop its blade and walk across the road route pushing the snow and debris off to the sides. Before the crews and log trucks move in, a road grader grades the route and fills in the holes with snow. On the Air Weapons Range that winter, Mistik used a water truck to freeze in the rough spots before loaded log trucks were allowed on to the freshly prepared roadbeds. However, on one wide corner, spring water bubbled up all winter, even when the air temperature dropped to -40 deg. C. However, think in reverse order when the melt starts and the frost leaves the ground – a bloody mess.

On February 18, Mistik finally gave the OK to move logging equipment on to the Air Weapons Range. However, by this time, all *genuine* logging operators were busy elsewhere. Not only that, but Silviba's road layout crew reported that Mistik was logging the ProForest timber permit from the south side. The Canoe Lake Indians had waited too long to give ProForest permission to use mechanical logging equipment. The only reason they finally succumbed, was that Jackie Ross spoke to Chief Guy Lariviere in Cree and 'lowered the boom'. I will remember that phone call from Mistik's office in Meadow Lake until my dying day. And, on March 3, Saskatchewan Environment & Resource Management at Meadow Lake [60] pulled everyone off the Air Weapons Range. ProForest had logged for just seven days out of a possible 65-day season.

Ω

On February 28, Sasktel cut off my home phone – destitution on its way?

Meanwhile, back at the Silviba office, Stuart North, the company administrator, kept going. Markus Klenz, a forestry graduate from Saskatchewan Institute of Applied Science and Technologies, had taught himself Geographic Information Systems. He was a genius. Eddie Kwiatkowski, another graduate, ran the Silviba field contracts. By March, several contracts were on the books. These involved Weyerhaeuser forest regeneration surveys and a Saskatchewan Environment & Resource Management plantation assessment project. Other contracts arose including from Prince Albert Model Forest silviculture research projects, and a Canadian Forces vegetation management plan at Dundurn. By this time, I was driving a ProForest Dodge 2500 diesel pickup – white. Weyerhaeuser was in close contact with Silviba as it pondered over the changeover of its GIS to another software platform. This, of course, would have serious ramifications for Silviba if it lost its ability to communicate with a major client.

Silviba's staff was exceptional in every way. When cashflow dried up, everyone shared in the pain. Stuart, who signed the cheques, canvassed everyone to decide who had needs on payday – this was usually rent payment, food and credit card bills. Stuart issued the balance of the pay cheques once more revenue arrived. The rule was that I was paid last. I cannot speak enough of the dedication and commitment by that crew. Eddie and Trent did all the ProForest logging layout and road location.

Ω

The preoccupation with economic development by the Northwest communities never ended. I spent much time at Meadow Lake or Buffalo Narrows on the various projects. All of this coalesced with work I did for NBC, but I have no recollection of what NBC stands for. Eventually, NBC, the Métis side of the proposed salvage through Louis Morin of Jans Bay, became a partner on the Air Weapons Range salvage. It crashed alongside ProForest from the Mistik/Saskatchewan Environment & Resource Management machinations.

Ω

Dundurn turned out to be a fascinating project. Canadian Forces used this range to store ammunition as well as train infantry and tank crews. Silviba received help with vegetative ecology mapping from

Wayne Harris, an independent biologist with his own company. The problem at Dundurn was twofold: tanks chewed the hell out of sensitive dune lands, and they shot live ammunition, including phosphorus shells that caused wild fires. This resulted in reduced vegetation cover. The danger of erosion grew exponentially.

So, a second part of the project was to review the standing fire orders for the base and revise the weak ones. I worked with the warrant officer i/c fire protection to upgrade a vegetation -based fire danger rating system. Since wind, combined with low relative humidity is lethal, the fire orders incorporated 'no smoking' zones as well. This was a fascinating contract. Wayne Harris, a consulting biologist, was a good teacher.

Ω

ProForest (4)

ProForest delivered its tiny volume of Air Weapons Range timber salvage to NorSask at Meadow Lake before the March 5 deadline. The strategizing over investment, partners, sawmill options, involving Carrier Lumber from Prince George and markets for forest products continued unabated. PFP's lumber broker, Al Verigin, reported the lumber market was strengthening significantly. This meant that higher cost logs also became economical. Air Weapons Range timber sources grew more attractive, although, burned trees have a salvage life of no more than three years in Saskatchewan. After this period, the wood is too dried out, contains cracks, and is infested with wood borers. Also, the jack pine displays blue stain.

A new business strategy emerged: franchising. The hewsaw concept triggered the idea of forming a franchise company. ProForest franchisees would set themselves up to produce lumber that they would then sell to ProForest for marketing. The ProForest board launched an investigation into potential timber sources, franchisees and investment sources. Not to forget the need for potential partners. Once again, the vibrations throughout the ProForest protoplasm awoke the good spirits. On April 29, Paul Parmenter from Stillwater Forest Products Ltd, based in Kalispell, Montana, arrived in Prince Albert to explore the ProForest franchising concept. Originally, Al

Verigin had contacted him to see if there was an interest in expanding Stillwater's marketing capacity.

On May 5, board members flew down to meet the Parmenter family in Kalispell, tour the sawmill, drool over Stillwater's finger jointing operation[61], and talk investment into launching a sawmill franchise company.

I toured the Air Weapons Range in a helicopter with yet another potential investor on June 19. By this time, I knew every air target location on the Air Weapons Range east of Primrose Lake. Jackie Ross' loggers had uncovered one bomb while logging. The authorities blew it up.

Ω

Ed Kwiatkowski graduated from Saskatchewan Institute of Applied Science and Technology, Prince Albert on June 22 after submitting his final assignment. He and Brenda were married on a gorgeous day in Tisdale on July 23.

Towards the end of July, ProForest directors directed me to advertise for a senior forester for forest administration duties on the Air Weapons Range.

Ω

Meanwhile, I prepared a briefing document for Ron Ray[62], the appointed Indian representative for a government-sponsored business development trip to China. Also, I accompanied Alphonse Bird on a business trip to Edmonton for some reason. I ended spending the next day at the federal Northern Forest Research Centre, with a representative from Alberta Environment and Lands. Both Alphonse Bird [63]and I had some common behavioral characteristics. We are both 'blue sky' thinkers. However, I felt more comfortable working from the bottom up, whereas Alphonse preferred the reverse. Thus, it could have been one of the latter's big ideas in the works. But what?

Ω

Silviba was busy with contracts. The design of an evaluation of the federal agroforestry initiative took up much time. By now, Markus must have replaced PAMAP GIS software with the new forest industry standard: ArcInfo from Esri.[64] Silviba would remain competitive. But who paid the $25,000 for the ArcInfo GIS licence, or did Silviba start off using a more basic version first? Most probably, the latter it was paid for through cashflow.

Ω

Silviba business plan development

A buzz of fieldwork, Prince Albert Model Forest projects, Indian mentoring, ProForest meetings with Mistik and Saskatchewan Environment & Resource Management continued. On September 23, someone told Silviba its services on the Air Weapons Range were no longer wanted. ProForest was now operating as ProForest Resource Managers Inc.. The Parmenters were now sliding into bed with Gary Benson through SGGF.[65] I spent the next three months on marketing trips in the Silviba red ford ranger. My ProForest white dodge was back at the dealer:

On October 24 I travelled to Lac La Biche to scout out the demand for Silviba forestry services. I met with Alberta Forest Service foresters to get a feel for consulting opportunities in that region (Alberta side of the Saskatchewan border). From there I travelled to Slave Lake and interviewed forestry staff at Slave Lake Pulp, Weyerhaeuser OSB, Tolko Industries and Vanderwall Contractors. I felt that, with perseverance, Silviba would get work there.

The trigger for these marketing travels was that Silviba needed to update its own business plan. I set an annual revenue target of $600,000 a year, on the low side, to $1.1 million on the high side. I had to document the likelihood of consulting contracts.

I was also looking into the possibility of a buyout from Geospatial International Inc., with headquarters in Burlington, Ontario. I started my second marketing journey on November 7. Once again, I visited the forest industries at Slave Lake, Alberta. My research pointed to a missed opportunity the first time out: Alberta-Pacific Forest Industries Inc. at Boyle, Alberta.

I travelled to Grande Prairie and Dawson Creek to introduce Silviba to Canfor, Ainsworth, and Louisiana Pacific timberland departments. On my way to Edmonton on November 10, I dropped in at Millar Western in Whitecourt, Alberta.

Ω

I flew to Burlington, Ontario on November 11 and spent three days visiting with Greg Wickware, the man at the helm of Geospatial International Inc. Greg gave me a tour of the facility. We discussed the mutuality of integrating the two companies, Geospatial and Silviba, and explored the options for a merger, buy out, or coalition. The visit went well. However, sometime later, Geospatial ended with an office in Prince Albert and a merger with Timberline Inventory Services of Edmonton. Since I had also approached Timberline, to explore similar scenarios to the ones discussed with Greg, I had to take the credit, or the blame, for bringing those two companies together. Later, I watched them struggling for survival, since I knew exactly what confronted them – too much overhead with insufficient revenue.

On November 18, I was back in Alberta, this time visiting forest companies in Drayton Valley, Grande Cache and Hinton. At Hinton, I interviewed the only forester to answer the ProForest advertisement for a senior forester. Eventually Rob Stanners did visit Prince Albert to explore the opportunity, but he went home not impressed.

From Hinton, I travelled up the back road from through Grande Cache to Grande Prairie and Valleyview. I had determined that Valleyview was the hub of all potential forestry work from Whitecourt in the south, to Grande Prairie in the west, then east to High Prairie and Slave Lake. I rented an apartment and hooked up an answering service before returning home.

Shortly after, Millar Western asked me to do a private woodlot assessment survey between Drayton Valley and Whitecourt. However, I had to return to Prince Albert meantime to finish off a hardwood supply study and contracts with the Prince Albert Model Forest.

261

The direct journey from Prince Albert to Valleyview is 960 km. I arrived back there on December 1, stayed at the apartment, found an old desk. The next day I returned to Millar Western in Whitecourt to pick up maps and the terms of reference for the private woodlot assessment bid. I spent December 3 at the apartment – it was so cold that I hunkered down early.

At about 0120 hours on Monday morning, lying on a foamy in my sleeping bag, I absorbed the presence of my dad joking and laughing as he stood at the end of the bed. I joined in the laughter. Pop's visit became part of my dream fabric that night.

The next morning, I breakfasted at the Husky service station up on the highway, and then phoned the office in Prince Albert to get messages. Stuart told me that my dad had passed away in the night in the Port Coquitlam hospital after a fall. I went back to the apartment, threw all my clothes into the back seat of the red ford ranger, and then pointed the way to Vancouver. I started off nearly blinded by tears, ran into glare ice between Whitecourt and Edson, and then turned on to the Yellowhead highway. I spent the night in Kamloops before arriving at the family 'ranch' on Eastern Drive in Port Coquitlam.

Clearwater Forest Products Ltd

I stopped in Meadow Lake on my way back from Vancouver to Prince Albert. The Saskatchewan Government Growth Fund (SGGF) and the Parmenters had joined forces to form Clearwater Forest Products, based in Meadow Lake. Parmenters were to supply a hewsaw, while Clearwater's logging department was under the management of Rod Sexsmith. Rod had run the logging arm of Weyerhaeuser's pulp and sawtimber supply operations. He took over the Air Weapons Range operations for Clearwater Forest Products. ProForest became a very limited partner in a limited partnership controlled by SGGF. Seemingly, Gary Benson (or Baron Benson as he was called) did not like ProForest. Someone asked me to give Jim Dawley, the Clearwater woods supervisor who answered to Rod Sexsmith, a familiarization trip on the Air Weapons Range.

Silviba ended with a contract to undertake some of the cutplan layout, ribbon in the access roads and take care of the bush scalers. Duties included the supervision of the ProForest cadre of small logging contractors hired to log on the Air Weapons Range. To get this contract, Silviba had to reduce its contract rate.

Chapter 24:

Silviba, 1995-96

Silviba's cashflow came in lumps; stretches of starvation before a glut of bill paying. On January 16 I noted in my diary: *'I will not put up with this starvation state again'*. I started the year with a big contract from the Canadian Forest Service in Winnipeg to review and evaluate all federal dollars spent on forestry projects in Manitoba over the last five years. Dr. Hugh Walker, an independent economist living in Saskatoon, worked with me to round out the economic side to the project evaluations.

The Manitoba Forest Service relied on its Hadashville forest nursery to supply most of the tree seedlings for outplanting into logged lands to uphold sustainable forest standards. The 1990 Manitoba Forest Resource Development Agreement set aside funds for building a new provincial forest nursery at The Pas. After running briefly, the provincial government locked the doors and retreated to the Hadashville nursery. Ye gods!

Hugh and I incorporated a massive supply of tact and hubris as we measured the millions of federal dollars spent to get this new nursery set up, before tearing it down to save running costs for the Manitoba taxpayers. The real reason was the technology for growing forest seedlings had changed. The evaluation report required us to include this fiasco, however delicately portrayed. To me this was not a new phenomenon. The Indian world had its own examples, hockey rinks especially. Indian Affairs provided the funds and the band built the rink. This included the installation of sophisticated ice-making equipment. Then reality struck since no one had included an operating budget for ongoing maintenance. Over time, vehicles drove through the doors, ice-making equipment destroyed, lights smashed and seats torn apart. This is not just an Indian behaviour. I note that municipal government, wherever, gets sucked into receiving grant money for vote-getting projects, then is required to find operating dollars from tax rolls.

Back to Manitoba. A forest nursery operated by a Dakota Sioux reserve at Portage la Prairie, and staffed by women band members, offered a model of due care and attention in growing tree seedlings for forestry use. The Manitoba government was encouraging this community to increase forest tree seedling growing capacity. This was 'bottom up' forestry, and so effective. Although it was small scale.

Ω

Through Prince Albert Model Forest, Silviba landed a number of contracts and subcontracts. A research contract with Prince Albert National Park produced three months of field survey and soil sampling for one three-person crew. Silviba was exceptionally fortunate to have folk who worked hard, were accurate with data collection and who could work independently. Markus Klenz produced the GIS maps despite computer and plotter glitches.

Gene Kimbley picked up unhappiness with Silviba work from the staff of the Prince Albert Model Forest on occasion: Geno: "I guess we have the wrong consultant"(re green draft of a map that was in dispute). He was constantly displaying signs of insecurity, and most of it was political (Métis vs treaty Indian). However, Silviba's technological insuffiencies occasionally created a feeling of *angst* within Silviba staff as well.

Small operators ordered ongoing cutplans and field layout services. These needed regular contacts and approvals from Forestry Branch. I turned these responsibilities over to Ed Kwiatkowski. The company administrator, Stuart North, also contributed mightily to Silviba's stability. My diaries suggest that Stu went out as a field-crew person at times.

Silviba did a traditional land mapping job for Montreal Lake Cree Nation that included interviews with elders. On these occasions, tobacco was passed around and smoked. I was fascinated by the definition of 'elder' in the Indian world. A few, old folk remembered the history of the people, especially with land use. Others were still in their forties as they sat looking up at maps and described their trapline, fishing and hunting locations. It was all so generic. Even the women were vague about the locations of berry picking. One understood that historically these folk kept any information to do with foraging to themselves. In gathering data, our investigations started with data from the 1950s; although two old men had an uncommon depth of knowledge. One restraint experienced by Silviba with this project was the elders could not (or would not) read a map.

Ω

On May 1, Silviba hired Kathy Michel, a Peter Ballantyne Cree Nation band member from Pelican Narrows as a GIS trainee. On May 8, Barry DeVocht phoned the Kordybans in Prince George to report massive forest fires in Saskatchewan.

The 1995 forest fire season was extremely difficult in Saskatchewan. Smoke and flames caused 16 evacuations forcing more than 2,000 people from their homes. Fires closed more than 20 highways, disrupted power grids, burned large tracts of valuable commercial forest. Fire bans were common and property loss substantial. Internal and external reviews of the province's fire program recommended changes to improve the way fire crews responded to wildfires. However, the reviews also suggested the province's old fire management policy was no longer valid.

On May 17, Rod Sexsmith, the timberlands manager for Clearwater Forest Products (CFP) asked Silviba to scout out many large new fires to discover timber salvage potential. Much internal confusion developed between Provincial Forest Products Ltd owned by Barry DeVocht, Jackie Ross, logger, and ProForest directors in Saskatoon. Competition for timber from Clearwater outside the Air Weapons Range resulted in some discordance. Barry and Jackie Ross were still scouting salvage opportunities from the 1994 fires north of Pinehouse and La Ronge. Barry even had Silviba submitting cutplans on his behalf to Forestry Branch. Both the Pinehouse community and PBCN Deschambault wished a working relationship with Clearwater. Then Jackie Ross separated from Provincial Forest Products to preserve his flexibility to go after his own logging contracts. Finally, a second cash call from SGGF (Saskatchewan Government Growth Fund) erased ProForest ambitions.

I had never heard of the term 'cash call'. Saskatchewan Government Growth Fund, as the controlling shareholder in Clearwater Forest Products Ltd, needed working capital for its operations. The Parmenters contributed the hewsaw set up at Meadow Lake. The sawmill operation contract became their responsibility. ProForest was left with less than five percent of the equity after another Saskatchewan Government Growth Fund cash call. My share of this was two percent. So much for Air Weapons Range salvage opportunities.

The Clearwater hewsaw, bought by the Parmenters as their contribution to the Clearwater limited partnership, installed in Meadow Lake at the place chosen by ProForest early on, did not run properly. Wood chips from burned logs were dumped in a landfill. The Parmenters were unhappy, both with the hewsaw and their experience with Saskatchewan Government Growth Fund, and Baron Benson. Some felt that Benson was double-dipping from both his jobs: SGGF and president of Clearwater Forest Products.

Ω

On June 17, the Kordybans returned to Saskatchewan to investigate the latest timber salvage opportunities. Silviba scouted potential timber salvage blocks for Rod Sexsmith. Aerial survey combined with road layout and cutplan submissions took up much staff time. One notes that most of these large 1995 fires occurred within the Weyerhaeuser Forest Management Agreement (FMA) north of Prince Albert.

Weyerhaeuser sourced wood chips from sawmills as wood supply for the pulpmill. It held up the timber salvage to find out how to keep char from salvage logging out of the wood chip supply. Forestry Branch held up all salvage permits until Weyerhaeuser had decided on the areas to log. On June 19, Saskatchewan Environment & Resource Management cancelled logging in green timber. Fire-killed timber took preference until the completion of the timber salvage. Eventually Weyerhaeuser presented plans to limit its salvage to root-burned trees only. This removed most of the risk of contamination of wood chips from char.

For some time Barry DeVocht had agonized intensely over whether he was working for ProForest or his own company, PFP. He had offered me a job working for him. So had Montreal Lake Cree Nation and Peter Ballantyne band. I burned up much invoiceable time helping Barry with his strategizing. His nephew, Dwight and wife, Wendy, moved into an office right beside Silviba's premises. Dwight took over lumber sales. Wendy became Barry's office manager. So Silviba now had an office neighbour, Provincial Forest Products Ltd. Wendy became a real asset as the Silviba workload intensified. She prepared cutplan maps for timber salvage applications.

We had an in-house day care set up for newly arrived Laura Lillian. I baby sat when Wendy had deadlines to meet for the preparation of cut plans. We were still colouring maps by hand in those days.

Barry's heart was in the right place, however, as were his aspirations. I continued to support him and provide what advice I could. Remember, if Barry successfully grew his business, Silviba also became busier.

The Carrier and Kordybans period (2)
On August 21, the Kordybans arrived from Prince George in their Beech King Air 300 airplane, complete with Terry Kuzma (TK), the Carrier Forest Products woodlands manager. By this time, Silviba had laid out road systems in the Late fire salvage area north of Prince Albert, and sent the cutplans to the government for Clearwater (Rod Sexsmith). The Kordybans, including Barry DeVocht, Jackie Ross and I spent all-day flying nine fire salvage opportunities from one side of Saskatchewan to the other. I made sure all the necessary maps were on board. Bill Kordyban Senior directed the pilot of the Carrier Beech King Air as we strained to look out of the aircraft windows in the attempt to assess the likelihood of burn salvage. The pilot

manoeuvred over these fires by dropping down to tree top level and reducing air speed to a point the stall warning hooter began its raucous warning. Senior was in his element providing the instructions to the pilot. He stood straddling the aisle, legs apart, and feet securely anchored to the seat stanchions on either side. He leaned over the pilot's shoulders and issued his instructions as he grasped the headrest on either side of him. Dramatic navigation. Silviba got much work because of that inspection.

Back in Silviba's office, Senior, Bill Jr and Terry Kuzma queried me about Silviba's ability to undertake the large amount of fieldwork needed. This included the office work to prepare the cutplans and obtain government permissions. The relationship advanced rapidly as follow-up works commenced. I was deeply aware as the meeting progressed that they sat in a 1929 building with desiccated, threadbare carpet and plywood tables supported by sawhorses. The premises originally housed the Indian Affairs medical facility and dental office. It was the start of a stimulating association.

Ω

Then followed a transition from undertaking salvage cutplans for Clearwater to doing similar work for Carrier Forest Products Ltd. Terry Kuzma instructed Silviba to prepare 20,000 cubic metre cutplans on both the Late and Tippo fires. The latter on the Besnard road some 240 km north, then west, of Prince Albert. It was closer to La Ronge; it was huge. The Saskatchewan Environment & Resource Management La Ronge office approved cutplans for Tippo. Just west of this fire, the Musqua fire also contained salvage timber. This was close to the Métis community of Pinehouse. These folk showed great interest in joining in the fire salvage.

On September 23, Barry, Jackie Ross and I went north to scout 1994 fires on the Pinehouse road on the way to Key Lake. Although we found good pockets of salvage wood, there was not enough volume to justify the cost of hauling logs all the way back to Prince Albert, almost 500 km.

On our way back home that afternoon, we visited the Tippo fire on the Besnard road (Highway 165) and checked out a possible sawmill site for one of Carrier's bush sawmill units. After that, as we drove past the junction where the Halfway road met the main road (Highway 2 North), we drove up into the Late fire to inspect a second sawmill site for Carrier near Pear Lake. I was revved up at the thought of a Carrier mill located at the fire site. Access to these fires was easy, since old logging roads crisscrossed the areas of interest. Again, I was aware of Senior's genius behind fabricating these bush mills. I love to work for people with 'genius'. Each mill was to arrive from Prince George in pieces mounted on semitrailers; seventeen of them welded together, with their own power plant and waste burners. Both the Late and Tippo fires had their own millsites.

Silviba's role expanded exponentially. We had ribboned out all the cutting boundaries and main roads for both fires by September 28. Clearwater released areas it was not interested in; Weyerhaeuser released areas too badly burned.

Carrier's log utilisation limit on the low end was a tree that would buck out a sixteen-foot log with a minimum top diameter of 4 inches. This gave it the edge over Clearwater who would not take a log less than sixteen feet long to a six-inch top. The reasons for this dichotomy are that Clearwater had to haul loaded log trucks up to 120 km to its hewsaw plant at Meadow Lake. Thus, it could only afford to take out timber with a higher lumber recovery per log delivered. This meant that it could produce lumber and still earn a profit. Carrier, on the other hand had two advantages. First, the sawmill was only a spitting distance from its own log supply. Second, Carrier mills incorporated two log lines. First, a chipper-canter for small logs processed these at great speed. Second, larger logs fed into the dimension sawmill to produce 2X8 and 2X10 dimensions of lumber. Each mill could produce 10 million board feet each year.

On October 18, I sat down with key staff to organize job descriptions for the Carrier Tippo and Carrier Late operations. Senior bought Silviba an HP map plotter for GIS use, then subtracted a portion of the cost from each Silviba-Carrier invoice. Eddie Kwiatkowski took over road layout and construction on the Late fire, while I made sure the Tippo loggers had road built ahead and need not have to stop operations. I flew to Vancouver on October 23. My mum had passed.

Jackie Ross had the difficult job of finding genuine loggers, who were now scarce. Eventually, he directed wood flow into the log storage yards on both fires. Local loggers leased used equipment from Carrier's bone yard of used logging equipment in Prince George. Winter was now approaching fast, the temperature dropped and it was time to make roads.

Carrier sent forester Joe Schvenkel from Prince George to be Terry Kuzma's man on the ground. This was November 7. I liaised with Saskatchewan Environment and Carrier to get approval for two sawmill waste burners; one for Tippo and one for Late. Hamish Duncan, an economic development guru based in the West, helped with gaining approvals. Saskatchewan Environment had firm specifications on air quality and rigidly enforced the air quality standards. These were the cleanest sawmill waste burners ever erected in the west at that time.

I remember discovering a flatbed semi loaded down with a D7 cat parked at the side of the road at Redwing on my way home. This is a small conclave sustaining a grocery and confectionery, gas station and the regional Polaris snowmobile dealer, just north of Prince Albert. It was a Saturday. I queried the truck driver who told me that he was heading to the Tippo fire to unload the cat. I gave him directions, and then went to look for Jackie Ross to tell him the cat had arrived. We joined forces and met the lowbed up on the Tippo fire. Jackie got up on the machine to start it so he could walk it off the lowbed. However, it was dead in the water. I have no memory of how we pulled that cat off the lowbed – Jackie must have pulled it off with a log skidder. Anyway, the trucker could return to Prince George to bring back another machine. The problem? Prince George had sent the cat out with summer diesel fuel – now that it had turned cold, the fuel had turned to jelly. I remember that Sunday morning as the first cold day of that season.

That was November 7. That same day, the Kordybans flew in from Prince George to tour the millsites and the bush from a helicopter. Bill Kordyban Sr called the shots, from laying out the bunkhouse locations and sewer lines, to the placement of log storage areas and weight scale shacks. This was all done with great precision. Together with Joe, Senior left behind some key workers to prepare the two sites, complete with bunkhouses, staff accommodation and weight scaling stations.

On Nov 27, Terry Kuzma told me to stay in the office, but given the nature of the startup of operations, this was not a relevant instruction. On November 29, Jackie Ross, a forestry assistant and I stayed in a small camping trailer. It was so cold, the condensation turned to ice in the doorway. We had turned on the burners on the cook stove. After the propane ran out, we could not close the door. With a brilliant moon casting shadows on the snow, I staggered out on one occasion to do my business. Those folk who have experienced dropping their pants at minus 42 deg. C will never forget it either.

On December 7, Jackie and I reconnoitred an ice crossing over the Smoothstone River. This would provide access to the north side, a part of the Tippo cutplan. After approval from Saskatchewan Environment & Resource Management fisheries people, Jackie set up his dad, Sandy Ross to build an ice bridge at that location. First, they strung logs across the river and tied these back to stumps on the other side. It was about 25 metres wide at this point. The logs slowed down the flow of the river. Ice formed between the logs. Sandy then pumped water on to the growing ice bridge until it was wide enough and thick enough (2 metres) to take a loaded log truck.

The La Ronge forestry officer had approved a section of the Tippo fire for Brian Zelinski, who ran a sawmill there. Silviba ran the boundaries. It did the same for Barry DeVocht, also on the Tippo, as well as Tom Sanderson and Marvin Newfeldt. Clearwater (Rod Sexsmith) had grabbed a stretch of timber just across the main road from Carrier's Tippo mill, and designated a logger, Eugene Boyer. However, Clearwater had to haul this wood all the way to Meadow Lake hewsaw mill, a one-way distance of 224 km.

Carrier set up Philip Tinker and 'Peanuts' from Pinehouse to log another block of timber close to the Tippo mill. Senior rented them a feller-buncher and grapple skidder. Poor old Philip never did not get much wood decked. That tired iron from Senior's Prince George boneyard had something to do with it. On the other hand, you cannot just sit in the cab of the buncher and expect it to work independently.

A logger needs a mechanic who can maintain equipment. These were not available in Pinehouse. Also parts departments were so far away.

For the next three weeks, Silviba staff put together cutplans for the Muskwa and Nipekamew fires, including field reconnaissance and main road locations.

On December 18, I took part in a large Prince Albert Model Forest meeting in Waskesiu (Prince Albert National Park). I presented the results of forest research projects through a Silviba contract. I had a ferociously abscessed tooth on that occasion, which is why I remember the occasion so vividly.

On the way back from my last trip to Tippo prior to Christmas, I brought back Carrier's two key camp construction employees to Prince Albert, Kelly and Howard, so that they could travel home to Prince George on the Company plane. They rode in the Silviba grey suburban. In spite of a rear heater, the trip was devastatingly cold. The wind whistled in through the weather stripping and under the door panels. And Kelly wrapped himself up in my emergency sleeping bag. Not a night to forget.

My last 1995 job was to help Jackie Ross on December 31 lay out a mining claim on Jackie's trapline at Two Forks, just south of the Besnard road junction with Highway 2 North. His dad lived there in a log cabin. The snow was deep, and I should have worn snowshoes.

1996

January to June

January started out extremely cold. This delayed logging. Silviba had three Carrier fires in active salvage mode. We ran the weight scale shacks at both Tippo and Late. They ran 24/7. Daily reports of logs received went to Prince George. Silviba had a crew of 23, including Markus Klenz running the GIS, and Stuart North on administration.

Just after Christmas, the "Russians" arrived at Tippo to log for Carrier. Early on, Monte Updergrove, a Carrier supervisor from Alberta, came up to run the Tippo salvage. He replaced Joe. He became the link between Silviba, Tippo and Prince George. I set them up with their own block of timber and Silviba ran the boundaries and located landings for decking logs. Their home is a rural community somewhere in Alberta. On their way home that spring, one was killed on the highway.

Ω

Picker trucks from Prince George and Alberta did most of the log haul. Once at the log landing in the bush, the truck driver climbed up to a seat out in the open air and manipulated a set of tongs mounted on the end of a boom attached to the truck[66]. He picked up every stick of wood (tree-length) then placed each one between the stakes until the truck was fully loaded. This went on day and night until March 26, when Carrier closed down the haul due to spring break up. Those truckers never quit working. There must have been two operators per truck, each on a 12-hour shift. There, they sat, out in the open in the dead of winter, in full moonlight, as they loaded the trucks. Once back in the log yard at the millsite, the truck driver unloaded every one of those trees piece by piece. As each loaded truck approached the yard, it passed over the weight scale and the kilograms were recorded. A statistical sampling plan showed whether a truck travelled through to the dump, or had to unload at a designated log scaling area beside the scale shack. When this occurred, the scale shack operator, who held a valid Saskatchewan scaling licence, measured every one of the pieces laid out on the ground. Afterwards, the calculations produced a conversion factor that converted kilograms into cubic metres. The government used this cubic volume as a means of collecting salvage dues (salvage $rate/m3 times metric volume flowing through).

The sawmills were up and running and bunkhouses in place. Carrier staffed the Tippo mill with two crews: A shift from Pinehouse and B shift from La Ronge. The competition for lumber output between the two crews produced some memorable records. A fourth fire was now in play: the Nipekamew fire. Silviba bought a white Chev 2500, quad cab diesel pickup. This was Silviba's fifth vehicle. It became the workhorse machine, and eventually replaced the former 4x4 suburban driven to ratshit beforehand by the RCMP in Wetaskawin. I had bought it from the dad of an RCMP officer, who

307

switched the automatic transmission to manual, then had to sell it because he was about to lose a leg from smoking too much. In all of Silviba's active years, I drove an average of 50,000 km each year.

By now, SaskTel had set up Silviba with a Fleetnet telephone system. This used the SaskTel towers erected here and there. There were dead spots though and these became known to Silviba in time. One day in February I was on the Tippo fire doing a timber reconnaissance in the middle of nowhere. I stopped in a patch of sunlight on a bank overlooking a large, brilliantly bright muskeg. I took off my gloves, pulled the Fleetnet out of the back of my timber cruiser's vest, and phoned Terry Kuzma in Prince George. Afterwards, I found that the thumb holding down the voice button was frost bitten. It was June before the last signs of that frostbite disappeared.

Ω

The same drill was ongoing at the Late fire closer to Prince Albert. Ed Kwiatkowski acted as Silviba's deputy there. Between him and Darcy Begrand, they kept ahead of the loggers. One remarkable absence from activity connected to the Late fire salvage was any contribution from Montreal Lake Cree Nation, only one hour travel from the millsite. Carrier did appoint the Montreal Lake Band Enterprises contract supervisor, to the position of the Late fire operations superintendent. However, this was not a happy relationship. Eventually, he disappeared back to the reserve after pulling the Pear Creek bridge in spite of being told not to by Gary Zuk, the Saskatchewan Environment conservation officer. I had tried to encourage Enterprises to sign a salvage logging contract with Carrier. I leaned on my sidekick, Gene Kimbley as well. There was just no expression of interest, in spite of the fact the Late sawmill was located on Montreal Lake traditional lands. But, Kelly Romanchuk from Prince Albert did round up logging equipment.

Ω

As January came to an end, Saskatchewan Environment & Resource Management released more timber for logging, both at Late and Tippo. It was an ignorant game to play. They dogmatically refused to allow the salvage of root burned trees. The needles were still green so they must be still alive. For Silviba it added to the workload because every little patch of root burned timber had to be marked and excluded from the logger's grasp. The loggers could not see under the snow and did half the logging at night. What made matters worse, was that as spring arrived, seemingly live trees lost their needles. Or turned red. Dead. So the machines would have to back track at extra cost to revisit those patches of now dead trees. I spent much time selling the resource officers on the need to allow logging on an area basis, not leave spots all over. In British Columbia, forest officers recognised that a logger would need some green timber mixed in with the dead to take some risk out of the salvage recovery. In some ways, it was an attempt to calm the 'no clearcutting' lobby which raised its head at the Environment minister's desk.

Ω

Stuart North had the responsibility of staffing the scale shacks at both Tippo and Late, all part of his Silviba administration duties. Stu had drawn up a subcontract that stipulated the qualifications and duties required for scale shack duties. Everyone signed, including one man from Meath Park who farmed in the summer. However, we fired him on February 19 for breaking his subcontract, and Stu found a replacement. The ensuing problem?

The farmer claimed UI. The subcontractor needed to pay CPP and UI contributions. But, if the applicant reported as an 'employee' rather than a subcontractor, Silviba had to submit the contributions on his behalf. To be an employee, an applicant had to demonstrate that 75 per cent of his annual income, or more, was sole source – even though he represented himself as a farmer from Meath Park. The farmer did this, Silviba appealed strenuously through its lawyer, but lost. Thus, Silviba paid up and the farmer received his UI. A lesson learned.

Ω

Barry DeVocht, Provincial Forest Products Ltd (PFP) bought an abandoned airplane hangar on the Sweetgrass reserve. After rounding up $1 million in government loan guarantees and grants, he bought a

¼ section in the about-to-become RM of Buckland industrial park. I surveyed in the approximate corner posts and ribboned in the road access.

Barry then commissioned Nick Maika, a millwright and sawmill designer from Merritt, British Columbia, to design and build a medium size mill. Barry then installed it in the building re-erected on PFP's new site. Hamish Duncan organized the construction of the most environmentally friendly wastewood burner ever built in western Canada. My concern was that after the timber salvage ended, where was Barry going to get the 180,000 cubic metres annually of softwood (spruce) to feed the mill? Prior to the timber salvage, his green allocation amounted to some 750 cubic metres a year, or less.

As if that was not enough, Barry found an abandoned wood pellet plant close by. The concept of wood pellet production was excellent, but ten years before its time. Thus, my time doing business plans and marketing studies went nowhere.

The Honourable Eldon Lautermilch came up from Regina to cut the ribbon and plant a tree at the new sawmill site.

Was that the end of it? No way. Barry and his nephew bought a tree length Morbark fencepost production machine. They set it up in the new mill yard. It had its own roof. This project became a success over time.

Barry struggled against a starvation timber supply time after time after time. Yet the timely cashflow provided to Silviba by Provincial Forest Products Ltd was important to Silviba's survival.

Ω

The next few months covered a number of actions:

- Silviba and the landlord replaced walls, electrical wiring, lighting and carpet at 1304 Central Avenue, Prince Albert – Silviba's workroom/office.
- Markus and I went on a marketing trip to Nelson, Vancouver and Grande Prairie. Markus dropped off to

310

attend the national convention of GIS practitioners, GIS 96, while I updated my professional qualifications through attendance at two courses, also a large forest sustainability symposium in Edmonton. Through the networks created, I updated the Silviba business plan/marketing plan.

- The Prince Albert Model Forest research subcontracts surfaced from time to time. Stu supervised relations with Dr. Thomas Bowman, director. It was usual for them to give out a contract to a high-priced environmental consulting company with its PhDs, and expect the consultant to hire a subcontractor for fieldwork and data collection. This was worthwhile to Silviba most of the time since it broadened Silviba's knowledge base.

- Regeneration surveys and WESBOGY[67] measurements for Weyerhaeuser. The latter was an interagency research project to establish plots in cutover areas, with continuing annual remeasurements to track growth rates. There were three locations in Weyerhaeuser Saskatchewan territory. The purpose was to track forest growth and succession. The boundaries at each location were permanently marked, each tree numbered with a tree tag, and corner posts identified on the ground. The objective was to compare progress within planted areas as opposed to areas left unplanted. Since the height of each tree was recorded in addition to dbh[68], it was painstaking. This needed extreme accuracy. Occasionally, a current year measurement, when compared to the same measure in the previous year, would uncover an error in height or diameter. Even some numbered trees were missed from time to time.

- Gordy McCrae started work. He was a forestry grad from Saskatchewan Institute of Applied Science and Technologies, Prince Albert. With a farming background, Gord was both versatile with equipment and created solutions to problems on a regular basis. At the same time, Jennifer Studer, a forestry assistant up at

311

Tippo, resigned after a particularly cold spell. I had worried about possible damage to her lungs. But after she called out to me on Central Avenue in Prince Albert sometime later, looking radiantly beautiful, I ceased to have that nagging worry.

- There were ongoing computer hassles. With the large databases now in play through cutplan development and research, those computers had to communicate with each other. Also, Markus had achieved skills with GIS analysis which became an essential component to Silviba's business suite.

- Ongoing cutplans and timber permit applications for both small and salvage operators.

- The re-emergence of the proposed Peter Ballantyne Integrated Resource Management Agreement (IRMA) initiative -- this time spearheaded by Deschambault.

- A resources map for the Pinehouse comanagement board.

- An offer by Gene Kimbley, on behalf of Montreal Lake Cree Nation to buy Silviba Services Ltd for $1 down.

Ω

July to December

On July 9, Chief Noland Henderson of Montreal Lake Cree Nation requested a meeting with me at the Northside coffee shop part way up the road to Montreal Lake from Prince Albert. He informed me that the band wished to terminate its relationship with Silviba. This was followed up with a signal from Geno to say that he felt himself to be in an uncomfortable position.

The period through to the end of September was full of bush work connected with permanent sample plot remeasurements for Weyerhaeuser at East Trout Lake. Add regeneration surveys and timber surveys before winter salvage. While camping out on one job in August, the crew survived rainfall that totaled 406 mm according to the Prince Albert National Park rain gauge in Waskesiu. I spent a week on another Prince Albert Model Forest /Montreal Lake Cree Nation forestry training project. I drew a line on a map that showed an area next to the reserve, and split the students into five groups. They had a week to develop then apply for a cutting permit.

I provided the aerial photos (borrowed from Forestry Branch). I showed everyone how to look at the photos through a stereoscope (to provide a 3D view of the terrain). We identified the timber stands containing trees of commercial size. Finally, we proceeded on the next day to prepare them for calculating the timber volumes observed.

On the third day, each group, using different coloured flagging tape (ribbons), went into the field with their maps and photos and timber cruisers vests and lunch to flag in access roads to the timber stands identified.

On the fourth day, each crew calculated the cost of logging their chosen timber stands. I had provided the equipment lists and forecast production levels earlier.

On the fifth day, they stayed in the office to prepare the cutplans and complete the paperwork needed for a timber permit application. The week was an outstanding success. A few weeks later, Prince Albert Model Forest organized a lunch at Waskesiu and Thomas handed out plaques to the best timber permit application group. Once again, the innate ability of young Indians, male and female, to show comfort with outdoor forestry work impressed me greatly.

Before exiting from these Model Forest sponsored forest assistant training projects at Montreal Lake I have one more to note. One summer, with the help of Markus Klenz, we did a forest ecosystem mapping project. The students had to do a plant collection as part of the exercise. They did a superb job of both the mapping and the plant collection. I learned as much as the students.

$$\Omega$$

In October Markus installed Silviba's first 4-gig drive before setting up a map update project for Weyerhaeuser. Since its timberlands office was just down the road to the pulpmill, the Company received frequent visits from Silviba staff. Silviba even loaned out Kathy Michel to the Weyerhaeuser GIS unit for routine data entry and other work. The Company also asked Silviba to bid on a 60 maps update project. On October 23, I remember a feeling of terror. Silviba could no longer talk to the Weyerhaeuser GIS software. They now had a fully equipped ARC INFO installation, compared with Silviba's primitive ARC VIEW. Without that cashflow, Silviba would have to shut down eventually. Also, Dale Reid, Peter Ballantyne Cree Nation adviser and assistant to Chief Ronny resigned his position.

Ω

In late October, Silviba laid out Carrier's Nipekamew salvage operation. This was a big job since very little previous logging access was in place. One of my stories involves the arrival of a logging contractor from Williams Lake sent in by Senior. Before they arrived, I flagged in the access road for some distance. My route took me across a muskeg. It ended on a bank before the road route continued on. I tied the ribbons up high so that the cat operator could see the ribbons above the carpet of black spruce rabbit bush covering the muskeg. I had explained earlier how a muskeg is frozen in to allow loaded log trucks to traverse it safely. Well, the Williams Lake boys drove their D8 cat, just regular width pads, on to the muskeg and quickly sunk into the ooze. They complained to Senior back in Prince George, so I got an earful, second hand, from Senior. However, after I met them for the first time, and explained how to freeze in a muskeg, everyone relaxed and got on with the job of logging. Their equipment, except for the cat, was brand-new, including a brand-new John Deere feller-buncher. It was a treat to work with them. They even brought along a completely equipped mechanical shop. They had placed it in a freight van and it had its own diesel power unit. The contractor even fabricated hydraulic hoses complete with attachments. This saved much downtime.

Early in November, Silviba won a forest regeneration survey bid from Saskatchewan Environment & Resource Management, with a spring 1997 start up. On November 24, Silviba received a large line of credit, and Barry DeVocht suddenly woke up to the fact he needed logs. Dale Reid had a battle on his hands with Revenue Canada. By this time, Eddie Kwiatkowski was a Carrier employee based in Prince Albert. It was a bittersweet happening for me. On another cold November evening, after a day locating a new forest road, I had walked back through the moonlight to my truck to find that it would not start. However, the Williams Lake boys came back later to check on me. They started my truck for me, and then returned to their accommodation at Whelan Bay.

By this time, Carrier's entire salvage program was well under way after freeze up, and Senior started looking for a place in Prince Albert to relocate the Tippo mill. Everyone put in the day on December 20 before going home for Christmas.

Carrier's persistence in creating a manufacturing base in Saskatchewan has ended with a more stable forest economy, especially after the demise of Weyerhaeuser Canada.

Ω

Chapter 25:

Silviba's worst year, 1997

It needs saying that Markus Klenz was the Silviba unsung hero in 1997, followed by Gordy McRae, Jackie Ross and Monica, Silviba's unofficial director of finance at the Prince Albert Credit Union. This is the year that I got down on my knees every morning to ask God to give me the wisdom, energy, endurance and patience to continue on the Silviba trail. Every night, as I closed my eyes before sleep, I recognised the outline of Jesus in the bottom right hand corner of my "screen". I saw the Silviba trail proceeding interminably forward. It appeared to have no end. The need for survival intensified because of the approach of a deep chasm on that trail, technological obsolescence.

What I did not see was the damage to cashflow perpetrated by some unscrupulous associates as the year ground on.

Ongoing professional and technical services

Silviba joined in (received contracts) forest research studies sponsored by Prince Albert Model Forest, due partly to its indirect association with both Gene Kimbley and Montreal Lake Cree Nation. These involved forest ecosite determination and later management protocols. Markus was the backbone of the forest ecosite investigations. These needed soils and vegetation classification and mapping as well as determination of tree volumes by species and age class. This project eventually became part of forest ecological management guidelines for Saskatchewan; some Silviba plot data originated in the Prince Albert National Park. These guidelines later formed part of preharvest silviculture prescription for all proposed

317

logging plans in Saskatchewan. The preharvest prescription became a guide to postharvest forest renewal following logging.

Weyerhaeuser Timberlands

One day, I made a date with Steve Smith, Weyerhaeuser Timberlands vp, to sell him on a forest ecosite approach to sustainable forest management. It was a precious interlude. We started in the panabode boardroom, and then carried on as Steve walked us through the pulpmill log yard and environs. The meeting lasted all afternoon – no lunch. Although quitting time for Weyerhaeuser Timberlands staff was 4 pm, Steve and I did not finish the meeting until 6 pm. The mission, of course, was to add a forest ecosite layer to Weyerhaeuser's forest inventory database. The meeting was sufficiently collegial that I, at one point, slapped Steve on the back of his hand to accentuate a particular need.

Early in the year, Dave Lindenas from Timberlands approached Silviba to discuss possible methods for use in updating forest inventory maps. The problem: after road building and logging, how do you update the inventory maps without having to put people on the ground to GPS and record every metre of timber edge and road constructed? Markus Klenz, Silviba's map update specialist, proposed to send up a photographer in a fixed wing aircraft (single engine) to take a flightline of photos through the bottom hatch. Silviba would use Lee Atkinson, a local photographer with all the right camera focal lengths to get the pictures. Markus designed the flight line locations to cover the update areas needed for the pilot's use.

Finally, the skies cleared and flying began. Markus or I acted as navigators and liaison between the pilot and the photographer. We needed the first flight to calibrate the photographic equipment by testing flying height against airspeed. Lee's job was to decide lens opening settings and frames per second. The aerial part of the job did get done, but Markus had the responsibility of transferring the air photo results on to Weyerhaeuser maps, then updating the maps on the GIS. I remember the results as crude. A flightline of photos is supposed to overlap to produce stereographic images. However, the drift from the aircraft line of flight, together with changes in altitude created some difficulty in transferring image data on to the GIS database. That single-engine light aircraft was not a stable enough platform. Eventually, Hilke Cray's company based in Edmonton took over the job of flying update photos. Hilke's much larger aircraft provided a stable platform for the cameras mounted in the hatch at the bottom of the fuselage. Silviba's job was to produce flightline maps for Hilke from information provided by Weyerhaeuser.

Why the need for update photography to begin with? First, Weyerhaeuser's forest management agreement with the Saskatchewan government included maintaining the forest inventory. Second, the inventory was also the basis of modelling the sustainability of the forest biomass over time. The values incorporated had to be 'net'. To achieve this, the GIS software program converted all logged over lands back to year zero. Thus the modelled volumes did not overstate possible harvest levels (sustainable allowable cut).

Weyerhaeuser's spring reforestation surveys came with new rules: Weyerhaeuser chose the survey patterns, not Silviba (change in Weyerhaeuser staff).

The Hudson Bay field disaster

Saskatchewan Environment and Resource Management awarded Silviba a contract to do a plantation assessment survey east of Hudson Bay, almost on the Saskatchewan-Manitoba border. In fact, some logging had trespassed into Manitoba. Access to this work site in the Porcupine Hills started at the junction of Highway 3 and Woody Lake Road (no. 980) at Armit. Remember, this is the location of Gene Kimbley's first posting as a resource officer in the former Saskatchewan Department of Natural Resources.

As you travel south on 980, you climb up the Brockelbank Hill onto a plateau, complete with fire tower. It is rolling country, bisected by deep gullies and creeks interspersed with small lakes, ponds and wet areas. Logging has gone on here on since the early 1900s. The soils are deep clays, impassable in winter, and even 980 with its gravel surface becomes impassable during the spring break up. It is winter logging country only. Originally, these Porcupine Hills supported large tracts of large diameter white spruce. However, it is all gone now, replaced by carpets of aspen poplar 12 metres tall, mixed in with struggling remnants of white spruce plantations. The remnants of old winter logging roads wind their way through the terrain, interrupted by washed out bridges, embankments and creeks.

Being higher up, snow was heavier in the Porcupine Hills than elsewhere. The six-man crew moved into its camp more or less at the centre of the work area some distance south of Armit. It had to clear away wet snow to create a patch of ground for tents. There was no communication except for a payphone some 20 km south at a lodge. I arranged for the lodge to provide showers for the crew. Dave Dorintosh was the Saskatchewan Environment & Resource Management resource official responsible for seeing the contract went according to the terms of reference. Dave approved the sampling plan and did checks on the quality of data collected. The crew used to keep up their spirits in a wet camp by visualizing Dave lying back on his double bed at the motel in Swan River. After a steak supper and two beers, he watched a porn movie – all in our imagination of course.

The work schedule revolved around a ten-day in, four-day off routine. However, this rhythm fell apart quickly. People quit and quads broke down. Vehicles got stuck, and BrockelBank Hill became impassable after a spring snow storm. A project that Silviba considered a routine turned into a nightmare. One rental pickup came back to Prince Albert with a smashed back window and a badly dented tailgate. The rental quads were in no better shape. After Dave Dorintosh inspected one crew's results, he refused to accept the plot data recorded. First Markus, then I had to redo survey lines and perform 'catch up' to get the work schedule back on track, and the contract completed. Dave was very good about giving the completed contract his OK after all the work was completed and verified.

The job was a disaster. I only bid on the work to keep the crew busy, but that was in the previous November, and the workload had intensified. However, Monica at Prince Albert Credit Union drew comfort from seeing the projected gross revenue expected. Without Gordy McRae's mechanical skills on-site, those quads would have had even more down time. Silviba licked its wounds then continued on to the eventual catastrophe.

Weyotun Resource Management Inc.

It all started in the fall of 1996 when we did that map for the Métis village of Pinehouse Comanagement board. I made various trips to Pinehouse to collect what information was available, before Markus put the map together for them. This was a period where too many players were barking up too many trees. Pinehouse is 440 km northwest of Prince Albert, so this meant an 18-hour day at the best of times.

The Pinehouse aspirations grew to include forming a logging company to log studbolt material for Carrier's Tippo mill. I became the 'go between'. Bill Kordyban Sr treated the Indian side of the equation with scrupulous attention. He was prepared to rent any equipment needed and provide the market. Green studs (as opposed to burned wood) were an attraction. Greg Ross, the mayor of Pinehouse, with Glen McCallum from the same community became the main Pinehouse contacts. Greg worked for Highways on the grader, thus had good connections. Philip Tinker and Peanuts had logged for Carrier at Tippo. It was true that most people in Pinehouse were poor, lived on social assistance and received subsidized housing. It was the Métis side of an Indian world. Many of the residents were former band members from the northern bands, including Canoe Lake and La Ronge. They lived on Pinehouse Lake, which joined up with the Churchill River north of them. If you keep driving north from Pinehouse on Greg's beautifully graded road, you pass over the Churchill River Bridge. Eventually you reach Key Lake, one of Saskatchewan's uranium operations. The trucks carrying 'yellow cake' pass through Pinehouse travelling south. Their route skirts the edge of the village.

I only have a hazy recollection of how Silviba became involved with Tippo Forest Products. Somehow, Gordy McRae and I conducted timber survey and ribboning in access roads to the timber. Once we stayed overnight at Darsana Lake Lodge on the way to Beauval. I remember taking a helicopter into no-man's land somewhere south of the Musqua fire. I was then dumped off to ribbon in a road going south about 15 km on snowshoes. Somebody, possibly Gordy, picked me up from where I walked out of the bush on to the Besnard/Beauval road. It was a grueling trip. Presumably, after that,

Carrier built the road, Markus produced the map and I arranged the necessary timber permit through Karen Waters in La Ronge. This could have all happened in January 1997, wedged between timber salvage work, Prince Albert Model Forest research contracts and small operator cutplans.

This northwest region of Saskatchewan is a conflict zone of sorts because the western edge of the Weyerhaeuser FMA meets with a hazy Mistik FMA boundary. I do remember the poplar from that region went to the Meadow Lake pulpmill through Beauval. The community of Beauval to the southwest of Pinehouse had its own comanagement aspirations. Tippo Forest Products eventually faded away.

Ω

Early in the year, Markus and I decided that we needed a stable source of income to augment Silviba cashflow. Somehow, one option surfaced quickly. The former Pinehouse connections from working with Tippo Forest Products wanted to get into fencepost manufacture. Simultaneously, Green Tree Fencing Supplies Ltd (Keith Ross and Maurice Warren) based in Prince Albert needed a supply of fenceposts urgently.

Serendipitously, Pinehouse had approval to salvage fire-killed jackpine post material, Green Tree Fencing Supplies needed someone to supply them with fenceposts and Jackie Ross (no connection) was a 'born again' post logger besides his logging contracts with Carrier salvage. Jackie never had enough wood ahead of him to keep him from going squirly. Then Clearwater Forest Products through Rod Sexsmith with its Meadow Lake hewsaw contacted Pinehouse. Rod was also hungry for studs[69], not necessarily green. For all this, Silviba appeared as the focal point.

That February, somehow, Pinehouse started talking with Green Tree who then talked with Markus and me. Maurice Warren, Green Tree's marketing guy (Keith's brother-in-law) made the approach. I asked my lawyer, Mitch Holash, to attend a meeting in Pinehouse with me to discuss setting up a fencepost manufacturing unit at McLurg creek on the Besnard Road, Km 170. Mitch outlined the requirements for a joint venture. He included a suggested share distribution. After general agreement, we returned to Prince Albert so Mitch could type up a document.

The name of the new business put forward by Pinehouse was *Weyotun Resource Management Inc.* This meant 'white bear' in Cree. Anyway, Weyotun Resource Management came into being (*Weyotun*). I spent the rest of February undertaking a deep search for all the ingredients of a business plan:

- Cubic metre conversion = number of posts per m3 of logs
- Logging cost per cubic metre for post material
- Truck haul cost/m3 delivered to peeler site
- Post specs: size and length
- No. posts per m3 for each size and length class
- Bucking cost = how much to pay post cutters on a production basis
- Peeler lease and maintenance /m3 projected
- Prices received for posts fob by category (size and length)
- All ancillary costs:
  *Fuel on site
  *Crew expense = sustenance, transportation, WCB, benefits, safety gear
  *Overhead = supervision, management, communications, travel
- Chainsaw purchases, chain, oil, gas, sharpening files, spare chain
- Loader (lease vs purchase or rental)

I completed the business plan then shopped around for a bank and associated line of credit. The government economic development agencies in La Ronge turned us down – I had counted heavily on grant funds or a loan from this group. Glen McCallum from Pinehouse was with me when we presented the application. The Royal Bank and Bank of Montreal also declined our request. We did get accepted by the Prince Albert Credit Union. Clearwater agreed to let us use its road system built to haul out its share of the Tippo salvage.

In return, we had to sell any studbolt material too big for post making. And Karen Waters expedited *Weyotun's* timber permit application for salvage post material.

Ω

On March 3, 1997 *Weyotun* became legal. The joint venture partners signed the necessary papers and Markus invested $15,000 into *Weyotun*. Green Tree signed an agreement to buy all posts fob the peeler site at the negotiated prices. The work began immediately. Barry DeVocht sent up his D7 to clear a working yard at McLurg, Jackie started to log studbolt timber for Clearwater close by and the Pinehouse boys started to make posts.

Keith Ross brought up a leased Morbark fencepost peeler. Loads of treelength salvage jack pine became the beginning of a yard inventory. A crew in the bush felled the timber, and a crew in the yard bucked up the treelength into posts of the desired length and diameter. A Fordson tractor equipped with forks moved the post blanks to the peeler. There a peeler crew peeled and bundled the posts. After that, the crew piled them into a crib, and then strapped each bundle before the tractor transferred them to a growing deck of bundled posts ready for pickup by a Green Tree flat deck.

The McLurg fencepost manufacturing site was the best of any in northern Saskatchewan. The base was pure sand that compacted nicely. Spring break up was a non-event since there was no presence of clay soils to interfere with the movement of loaders, trucks or lighter traffic. It was an ideal year-round site. Before we were allowed on to the site, Karen told us to prepare an environmental assessment before site clearing. McLurg Creek did run through the eastern border, but we ribboned in a buffer to separate the riparian borders and creek from the operation.

All through March, we were short of loading capacity for hauling out both studbolt material logged by Jackie Ross as part of our Clearwater road use agreement and our fencepost tree length material. Markus must have found us a lease machine from Korpan in Saskatoon. One of the Pinehouse guys learned to operate it. Jackie's wood was loaded out. He then set up a bucking system in the *Weyotun*

324

yard. He rounded up a few skilled folk from La Ronge to augment the post production. Vic McCallum, Glen's brother from Pinehouse, learned to operate the peeler.

Then the hiccups started

Pinehouse terminated its joint venture relationship. Greg Ross wanted their $10,000 back, I was totally flummoxed. Why? I found out eventually that Clearwater was wooing Pinehouse away from us. The Company wanted to access Pinehouse 'traditional lands' and mop up all the logs suitable for lumber.

Green Tree was delivering our post material to Jennifer at her treatment plant in Dunmore, Alberta. On April 16, Jennifer contacted me to offer *Weyotun* an advance on deliveries. This seemed to be an attempt at an end run that would leave Green Tree as just the post hauler. Keith had cut us a cheque promptly for each load hauled out. On April 28, a massive snow shut down operations. Gordy McRae had kept the peeler serviced and functioning. Markus and I did not believe that Silviba could continue to function after May. The bills from the Hudson Bay fieldwork were higher than the revenues received.

Much conniving began between Peanuts (the Pinehouse logger complete with his own skidder and bunkhouse), Keith Ross of Greentree and Jennifer from Dunmore, Alberta. The leased peeler at McLurg needed a $2,500 hydraulic pump replacement. Thanks to Gordy we kept the peeler functioning. Then on May 26, Gordy took the peeler to the KLS mechanical shop in Prince Albert with engine problems. KLS found a secondhand replacement engine in Edmonton. I loaned Vic McCallum our red pickup and he drove there to pick it up.

By the end of May, the conniving to do an end run around me continued. I suppose that Jennifer was the instigator. The picture I can now paint contains Jennifer as the instigator since she resented Keith as a broker. But Keith, to keep his position with her, aimed to replace me as the supplier, and rounded up Peanuts as the logger so *Weyotun* could continue to ship direct to Jennifer. On June 4, Green Tree told me that they were 'pulling the pin'.

On June 6, I attended a Pinehouse Comanagement board meeting attended by Peanuts, Philip Tinker and Clearwater. This time, it was Clearwater trying to dislodge *Weyotun* in favour of studbolts rather than posts. We finally received the functioning peeler back on June 16. Pinehouse continued to agitate even while we continued with the post operation. Half the crew were Pinehouse people, the other half from La Ronge. Barry DeVocht's D7 cat built roads. Jackie Ross superintended the bucking of tree length into posts in our yard.

On July 7, Jennifer started to deal directly with *Weyotun,* she came up from Dunmore with her key men plus a loader and peeler. By this time, Keith Ross had retrieved his peeler, including the new hydraulic pump, in the dead of night. There were also rumours of folk stealing posts out of the McLurg yard. Jennifer continued to agonise over buying us out. She should have, if only to secure a good supply of good post material. We continued to ship her posts and she continued not to pay us until she had processed a truckload through her treatment plant. And we did this with nine more loads. Even Monica from the Prince Albert Credit Union phoned her office in Dunmore to explain that this arrangement would break *Weyotun.* All this while, I scaled up the manufactured post bundles as required by the Saskatchewan timber scaling regulations, ribboned in logging roads as needed. A condition of our timber permit was that we had to bypass burned timber stands that had a full stocking of new trees growing back. So forest regeneration surveys were another part of my job description.

Now, a new scenario was emerging fast. Somehow, John McLeod, an Indian and director of economic development for Waterhen First Nation (see Google) just north of Meadow Lake, found me and asked me to do a training package for his sawmill on his reserve. Since McLurg was a shortish distance north of the reserve, I travelled down and met John there. John was bullish about economic development. He showed me a row of houses with new roofs, and other reserve improvements that he completed. We toured his little sawmill and he briefed me on his expectations. We got on well.

We spent some time on a log sharing our experiences. John related how he had worked as an assistant to a heavy-duty recycling guru somewhere in California. This chap was obviously a mentor at some stage. He taught John that to succeed he had to kick ass. John also explained the White side of his family from Scotland was to show up in Winnipeg that summer for a visit. He had traced his origins back to the staff of the North West Co. fur traders back in the days of Rupert's Land and the Hudson Bay Co. I was fascinated.

We got around to 'Weyotun' and its work at McLurg with Pinehouse folk. I explained our relationship with Jennifer and that we had a future but the bank account was zero. This meant an imminent shut down. Before you know it, he arrived at McLurg with Waterhen Chief and council, toured the operation, and told me to draft up a business transfer letter for the band to sign. On August 3, we co-signed the letter and Markus and I no longer owned *Weyotun* officially, though there was the matter of Fleetnet radios to retrieve. Also, I had left our Silviba suburban at McLurg so the crew could travel to and fro from Pinehouse.

Before you know it, he rounded up a logger from Alberta called Kuevers. This outfit brought in a forwarder, log skidder and a high capacity loader. Jennifer's man supervised yard activities, including post peeling and manufacture. But John found that Jennifer continued the practice of not paying until each load was treated and stacked for sale. I suspect her dad had taught her not pay until she had a sales order in hand. Therefore, we financed her side of the post business as well.

On August 19, Monica from PACU directed me to close down the *Weyotun* account. I gave John the company seal.

On September 7, I wrote in my diary: 'The beginning of the most defining weeks in my Silviba life'. We had no payroll and bills kept pouring in. Carrier had asked me to do a proforma on the manufacture of fenceposts. I could do this from firsthand information and they agreed to meet John and explore common opportunities.

When John returned from his visit with the Scottish ancestors, he and Carrier had their meeting in my office. Bill Kordyban Sr, Terry Kuzma and Bill Kordyban Jr sat across from John. I was somewhere in between. John started the discussion by painting a picture of the mutuality of their needs. John wanted to be in the fencepost manufacture business but he needed a joint venture partner. They had an earnest discussion about the ways to put such a deal together. Terry Kuzma had my brief beside him. Then John upped the ante, started to become hostile before losing his temper.

I was embarrassed and depressed at the thought I had caused all this to happen. The recycling guru must have taught John some really poor business tactics; so that was the end of a relationship before it had even started. Heartbreak.

I still travelled to McLurg to scale up the bundles with my newly gained Saskatchewan timber scaler's licence. Jennifer requested further timber permits. Monica from Prince Albert Credit Union insisted on a further discussion with Jennifer. On September, Jennifer suggested that we ditch John and form a new company. This I did with help from Mitch, and we formed TJ Posts Ltd.

Ω

On September 21, John phoned and suggested that we go bankrupt and start over. Don't forget, the brunt of the growing catastrophe was born by Glim Holdings, my ProForest shareholder.

On top of later threats from John, garnishees from Revenue Canada and the threat of a sheriff's visit from the Saskatchewan Labour Board, we persevered. John refused to give up the Fleetnet telephones, so I travelled up to McLurg and found them on the cookhouse table. The only missing one was in John's care. Not only did John refuse to give it up, he threatened me within an inch of my life. I left the suburban at McLurg and never went back. We owed thousands on those phones. One big problem was that SaskTel had no way of cutting off the Fleetnet signal, so the bills continued to grow – frustrating.

Ω

On October 16, the Pinehouse crew declared personal bankruptcy. I arranged for a receiver in bankruptcy and helped them through the process. On October 25, I found myself at Jennifer's treatment plant in Dunmore. The treatment cylinder, floor and pipes were glistening clean. She had no post inventory. I met both Randy and Leonard, her key men.

She took me back to her home in Medicine Hat and told me her story. Her dad died and left her to run the business. She had her two key employees but lacked a secure source of posts. It helped that she sat on the Alberta Forest Permits Board so she could acquire the rights to a source of fencepost timber. However, it was all lodgepole pine. This may be taller and easier to peel than jackpine, but it does not have the strength of a jack pine post. It was a pleasant visit. She had asked me to draft a letter of recovery for *Weyotun*, and get her some more permits.

On December 1, Markus Klenz, righthand man, ecologist, GIS operator and map and database expert, disappeared on a sabbatical. In 1998, as you will see, I started to wrap up the pieces.

Amisk-Atik FMA (Peter Ballantyne Cree Nation)
A new phase surfaced in the struggle to get the Saskatchewan government to enter an agreement for a new integrated forest management entity covering Peter Ballantyne Cree Nation (PBCN) traditional lands. The band offices now moved into old residential buildings on its urban reserve in Prince Albert. This included the band's economic development unit headed by Trevor Ives, with Ron Ray as the band liaison with Chief and council.

The year ended with much more work for Carrier on the Nipekamew fire salvage. Silviba was never short of work, just payroll that had now become a huge iceberg in the sea of tranquility. I have mentioned the matter of Revenue Canada issuing a garnishee. I had to list Weyerhaeuser in Silviba's list of receivables, so this agency scooped monies owing by Weyerhaeuser to Silviba. Now, Weyerhaeuser got really upset when forced to hand over cash in this fashion. Accordingly, Silviba was reprimanded and warned against this occurring again.

The termination of Silviba's business relationship with Montreal Lake Cree Nation

Finally, at some time before year end, I was summoned to a Montreal Lake Cree Nation meeting of Chief and council at the Imperial 400 motel in Prince Albert. Council was angry and upset. I briefed them on Silviba's failing financial position. I told them that Silviba's digital mapping capacity was out-of-date and that I could not bid on further Weyerhaeuser or government work. Since I felt that bankruptcy could be part of a survival strategy, I put a loonie on the table and told them I had bought back their shares. They had not put a nickel into Silviba for seven years. I knew that under the Indian Act the band was isolated from attempts to have Silviba debt foisted on them. But, I did not wish to have someone chase them for Silviba debts.

Now, what unfurled in 1998?

Ω

Chapter 26:

Silviba, 1998

Background noise

While this accounting of Silviba transpired, it was accompanied by much background noise. This could zap energy from the system. Perhaps the major energy loss was that since Silviba's assets were knowledge-based, the bank considered this as an intangible asset, and thus not visible on a balance sheet. No assets, therefore no line of credit. What little cash advance was possible, was due to the list of accounts receivable. Although payroll dollars were now due from the job just invoiced, it could be as much as ten weeks before the cheque showed up in the mail.

Weyerhaeuser played this game back in the early 90s by dumping its cashflow problems on to its suppliers. Let the supplier become a creditor. Indians were the next biggest headache. Because of the nature of the cashflow, Indian cash from economic development envelopes came in fits and spurts, so Silviba was paid in fits and spurts. It was not deliberate, but the more primitive administration on reserve exacerbated the situation at times. What did burn up significant energy was chasing down cheque signers. This could mean a drive up to Pelican Narrows, 420 km one way to find the cheque signer was over at Southend. So, why didn't I phone ahead before I made the trip? Often, the cheque signer had left the reserve to go somewhere, and not told the office. Peter Ballantyne's administration matured once it was established on the urban reserve in Prince Albert. The money collection stress was reduced. Actually, I had developed a number of contacts in the band offices on the various reserves. They were all women, had immense senses of humour, and were always a joy to drop in on. Chief Ron Michel's general conduct of tolerance, humour and patience had much to do with this. Frequently, it was a real chore to collect an Indian cheque. This topic will come back into play a little later in this Silviba narrative.

The major component in operating a business containing staff is covering payroll. It is vital to healthy business relations. No one working for government has the slightest inkling of the stress level generated by the business owner or manager during cashflow crises. Also, that bugaboo, GST. The government required that GST be remitted every three months. This added to a cashflow crisis at times – actually, most of the time. Although, I must add that Indian invoices were free of GST as long as the service was performed on reserve.

On the other hand, small operators paid their invoices promptly. I enjoyed a close relationship with them. In some sense, they became family.

Of course, no cashflow hassle was complete without heat supplied by overdue trade accounts, and MasterCard. This latter instrument was a vital part of Silviba's informal 'line of credit'.

The next major drain on energy was vehicle breakdowns. Everything from engine replacement, differential repairs to routine maintenance of brakes, tires, windshields, back windows, oil changes and radiators. Those vehicles had to operate reliably. A crew coming out of the bush at 6 pm in February had to rely on a truck that would start at minus 36 deg. C. So, add battery health to the maintenance list. Then there was the ancillary equipment: chainsaws brush saws, ATVs and skidoos. All in tiptop condition. Have you ever loaded an ATV into the back of a pickup and then squeezed the accelerator to get the machine up the loading ramp? Then had the machine lurch on to the pickup box and have the front rack pierce the rear window before you apply the brake? During its Prince Albert sojourn, Silviba replaced four rear pickup windows over time.

Government contracts had the worst outcomes. They involved much data collection. The government checkers were so slow in approving the results. On one job, the checker sat on the results for six months, then rejected part of the work. By then it was midwinter, and the plots were under 2 feet of snow. She told her colleagues over coffee that it was her policy to screw all contractors. Silviba spent a year trying to get paid. Also, government loves to hang onto the holdback – another way of saving money.

There was always a fair amount of competition in the forestry consulting service arena. In its early years, Silviba had SaskTel add another heading to the Yellow Pages: 'Consulting Foresters'. It was amazing how speedily every Tom, Dick and Harry listed themselves. To reference yourself as a 'forester', you needed to hang a professional registration certificate on your office wall from a designated provincial foresters' association. However, newspaper reporters in Prince Albert had coined the term 'independent foresters' for application to anyone connected to forestry outside government. Even Weyerhaeuser was labelled as 'independent foresters' by one reporter. The sad point was that this was a sign of how ignorant Saskatchewan society had become in its understanding of the role of forestry within the economic environment.

Other parts of the general energy drain at times included poisoned relationships with clients and invasion of Silviba turf by outside consultants. Both Timberline Forest Inventory Services and Geospatial set up shop as a joint venture in Prince Albert. This merged company worked from 'the top down' so they schmoozed successfully with vice presidents, deputy ministers and Forestry Branch managers. They survived on large contracts worth millions. Silviba's business plan included one scenario that would bring in $600,000 gross revenue a year, but the median expectation was about $350,000.

On March 5, 1998, I received the first tranche, $30,000, from my parents' estate. I paid out all this to the Department of Labour payroll claims and Revenue Canada. This lessened the stress level.

Weyerhaeuser re-inventory and mapping of its Saskatchewan Forest Management Agreement area

In 1998, Weyerhaeuser issued the terms of reference (TOR) for a re-inventory of all forest lands within the forest management agreement area. Given that the gross area involved was around 96,000 square km, it was a massive job. The TOR was complex. Each inventory component required its own map layer:

- Soils classification, by moisture regimes
- Ground vegetation classification
- Shrub layers
- Tree canopy
  *Mapped by cover types
  *Total age of each timber stand
  *Total height, tip to toe
  *Tree diameter at breast height
- Riparian overlays
  *Creeks, rivers
  *Boundaries of non – tree shrubs around water
  *Mapping of edges around muskegs, swamps, wet areas
- Tree volume sampling, using prism plots randomly located throughout the map area and distributed according to the representation of forest cover types existing.

This Terms of Reference (TOR) was partly due to my daylong meeting with Steve Smith, vp of Weyerhaeuser Timberlands in Saskatchewan referenced earlier. In preparing the proposal, Silviba's first job was to meet in Saskatoon with a consortium of professionals in soils, vegetation, riparian environment and ecology. The TOR also needed access to a more robust GIS than Silviba's outmoded software. Not part of the TOR, but important, was that government specifications for forest inventory were to be a guideline to the design of the end product.

In Saskatoon, I met with a consortium of the best environmental consultants in Saskatchewan, or even in western Canada. Perhaps, the man that drew it altogether was John Polson with his expertise in digital imagery. There were biologists, botanists, ecologists and specialists in riparian ecology at the meeting, and during the ensuing project design. Within the time frame fixed within the TOR, the team produced a costed out project proposal that it submitted at $26,730 per mapsheet.

Timberline Inventory Services from Edmonton got the job at $9,500 per mapsheet. However, not before Weyerhaeuser hired Silviba directly to do a due diligence on the six bids submitted. Silviba's consortium bid was in the middle of the pack. The due diligence conclusion was that Timberline met all the needs. Part of the original bid package was to do as complete test mapsheet and associated database supplied by Weyerhaeuser Timberlands. Doing a due diligence on the work of the largest forest consultancies in western Canada was instructional.

Jason Nelson

Jason was a graduate of the soils and crops program at the University of Saskatchewan. I first met him at a planning session associated with the Weyerhaeuser bid – soils component. Jason and Dr. Dynes, an agricultural soils professor, had formed a company, Nelson-Dynes in Saskatoon. They were close to the airport, and above the FedEx offices – a strategic location.

Jason's qualities in both soils and ecology were really attractive to me, both professionally and from a business perspective. Since my forest soils training during my undergraduate years was pitiful, the new association with Jason brought 'soils' to life for the first time. It was the start of an intense relationship. Jason was Gen X; no career opportunity and stifled by lack of work in his field. He was a loner, had large disagreements with the professor overseeing his thesis (not Dr. Dynes) and was a very frustrated, impecunious young man. His wife worked in a clothing store, but money was in short supply. As a result, Jason had fits of depression and frustration. They lived in an apartment block not far from Robin's Donuts on 8th street in Saskatoon. Jason quit his relationship with Nelson-Dynes and became part of Silviba's stable of associates. Jason's alternate office was now Robin's Donuts and I met him there frequently. I got Jason involved in a couple of soils jobs, one at Dundurn military base and the other on the Sturgeon Lake Cree Nation reserve north of Prince Albert. For the first time Silviba could pick up projects requiring forest soils expertise and ecological site mapping.

Eventually, Jason and Sharon moved to Prince Albert to be part of Silviba's team. Although, he was anything but a team player. However, his forest ecological skills blossomed magnificently, and everyone learned from him. He should have returned to university to take a PhD, but he could not put up with the 'bull'. Jason was an ecological detective in reality. He linked associated climate, soils, ground vegetation and soil moisture regime and tree canopy components into an ecological envelope on any piece of forest land. Foresters had the basis for all the tools necessary for forest sustainability. Jason is a genius. His projects expanded into the Northwest Territories, Alberta and New Brunswick. The federal forest researchers hired him to give advice in problem areas. Silviba may not have got work on that Weyerhaeuser forest inventory, but it achieved the next best thing, Jason.

The Carrier projects continued (3)

In 1998 more forest fires burned bright, then created more timber salvage. These were convenient fires for the most part, since they were not far north – easy to get to. Silviba did the necessary cutplans, road layout and boundary ribboning and obtained the timber permits. Both Jackie Ross and Barry DeVocht took out fire salvage contracts, as did a few others. This brings me to:

Fred John

I do not remember how Fred entered Silviba's life, but he became my timber layout partner and salvage assistant. Fred was a treaty Indian from up north. He lived with his partner and a new baby on the north side of the Diefenbaker Bridge in that little community beside the North Saskatchewan River. He was an energetic young man who could think and solve problems as he went. He was comfortable in the bush and I remember one 14-hour day on the Nipekamew fire salvage layout. He was also useful in the office preparing cutplans and other chores. He was so easy to train.

But, Fred was an alcoholic. He had the disease big time. It was unfortunate. He was a capable field man. In my time with Fred, he was jailed twice and found in his Silviba pickup passed out twice. His drinking hole was the Northside restaurant/bar on the way to Montreal Lake. Fred was pure talent with a wasting disease. Later, whenever I enquired, the news re Fred was depressing.

Carrier's continuing establishment efforts (4)

The Carrier salvage layout work went well. At one stage, Terry Kuzma told me to reduce the Silviba invoice or he would refuse payment. Another one of these foreign interventions occurred next. Somehow I had brought Mark Hislop into play. After the first Prince Albert sawmill site application, the government insisted on a public meeting as part of the approval process. I got Mark to organize the meeting subject to Bill Kordyban Sr's approval of all the arrangements. Mark had a communications company in Prince Albert, and somehow, the attachment with Carrier grew until one day Mark relayed a message from Carrier: 'If you go out to the sawmill site, you won't get paid.'

I used Mark Hislop to relay messages as Carrier needed to find another sawmill site. The public had turned Carrier down – a housing subdivision across the street yelled loudly enough that municipal politicians had to join in the chorus of 'no's'.

After doing a recce of two more sites in the RM of Buckland close by our original choice, I chose the NE 28 one. This was a mile west of the pulpmill in vacant Crown land, and close to a main haul road and the pulpmill rail spur. Fortuitously, it was also next to the electric grid. The drainage was good and the site plenty large. It was the quarter section right beside Barry DeVocht's new millsite. This site, of course was subject to Senior's approval. Thus, on August 27, the Carrier senior management flew into Prince Albert. I picked them up and took them to the proposed site. I had surveyed in the proposed access roads on the weekend. Since my home was only three miles from the new sawmill site, Carrier got a lot of my time free of charge.

Silviba undertook the necessary environmental impact assessment. Saskatchewan Environment insisted on a long list of 'musts' before approving the operations plan. Fuel spill protection, a rare plant survey, noise mitigation and approval of the waste burner design and combustion efficiency were the prime environmental criteria. Carrier brought in the waste burner from the Tippo millsite and re-erected it. Saskatchewan Lands Branch also became involved, since the millsite was classed as vacant Crown land.

I must point out that working with Saskatchewan Environment staff was a treat to associate with. They knew their job and were knowledgeable with the file. I learned much from them in the process of obtaining environmental approvals.

Silviba organized the access roads, ran the boundaries and tied in the corner posts, then worked with the RM road foreman to build the north access of the pulpmill haul road. After all the timber was logged and hauled away within site boundaries flagged by Silviba, Senior hired a land surveyor to rerun the boundaries at government request. My west boundary was 9 metres outside the survey posts, a function of a 1998 GPS instrument's accuracy. Senior was relieved. The site was now accessible from two directions; it was convenient, had good access and was accessible to rail. One joy working for Carrier was the work of the senior cat operator, Jessie. He made it all took so easy. The millsite preparation, once finished, provided a perfect log yard with the access roads ready to drive on. The distance to Prince Albert downtown was eight km. Omnitrax built a short spur line into the sawmill site itself. Eventually Senior moved the Tippo mill on to the Prince Albert site. By 1998, the 1993 fires were producing logs with cracks and shake, so Carrier went after both green wood and 1998 fire-kill.

$$\Omega$$

In August, I took my grandson Matthew and dog, Rennie up to Besnard Lake. We hired a boatman and travelled to the far end of the lake. A great wind sent huge waves against us. Landing on the rocks at the other end was difficult. Anyway, we completed a timber survey of that end of the lake, before returning to the boat launch in calmer water. On my return to the office, I did the cutplan and sent it to Carrier. That is one job that Silviba did not get paid for.

Unser log storage site

Silviba surveyed a potential site on orders from Senior, between Unser Lake and Limestone Lake. This was some fifteen km east of the potential Deschambault road junction on the road to Flin Flon. (See Google Earth). Carrier needed an overflow log storage yard, or 'surge bin', to serve as an alternate log storage site at times of heavy winter logging. The site was part of the 1963 fire. Little remained but scrubby aspen on a gravel bed over rock.

Carrier would have shipped lumber from Deschambault. This company had successful joint ventures with Indian nations, both in Alberta and British Columbia. Carrier included an airstrip in its plans. Perhaps someone in the Deschambault fold glimpsed the possibility of their last sunset, and panicked at the thought of an efficient third party taking over. Essentially, they ran Carrier off the premises without acknowledging intensive investigation into wood supply. Anyway, I felt that years of persistent coaching, mentoring and planning were for naught. Jack Custer from Deschambault who stood with me as our vision expanded from community-based to band-based over the years, understood the lost opportunity also.

Henry and Jack – again
I worked on community-based forestry projects out of Pelican Narrows. As a member of the band council, and later as economic development contact for the band, Henry spent his time coordinating the wants of band members with funding agencies. I wrote up the applications, many as freebies. Silviba's invoicing for work connected to the band's forestry efforts was about 1/3 of the total Silviba effort expended. Henry contributed to much of the background buzz referenced earlier. I appreciated those fishing and skidoo trips with Henry Morin.

My relationship with Jack Custer was long lasting, and at times, intense. I served as the buffer between Indian Affairs and Custer's Deschambault forestry imperatives. Keith Chayter from Indian Affairs in Prince Albert was the link between Jack's forestry ambitions on reserve forestlands around Deschambault and the needs of the Indian Forest regulations. Keith was a forester, and a rare breed of individual within Indian Affairs, but he did not last there. Silviba was the buffer between Jack's timber aspirations and forest management requirements: stream protection, tree planting after logging, and clean logging (no abandoned logs afterwards).

Ω

Through Jack and Dale Reid, I kept a close relationship with Deschambault over the years and well into the FMA planning stage. After Alan Appleby entered the picture it was time for Silviba and me to fade away.

Alan Appleby, through the World Wildlife Fund, did award Silviba a contract to map the Suggi wetlands south of Deschambault using satellite imagery. Markus Klenz did the pioneer work to classify the images and then transfer the results on to hard copy maps.

Ω

Gene Kimbley approached me at the end of December to hint at some form of business association. He had become self-employed since September. We continued to explore the options further in 1999.

Ω

Chapter 27:

Silviba, 1999

The year started off with the usual Silviba routines, except that two new themes though separate, knit the year together. Fred and I still dashed out to flag new salvage access roads for the loggers, and the 1998 forest fires continued to provide logs to small operators. Most of these arrived in Carrier's log yard. On January 25, I received a further $48,000 from my parents' estate. By May 7, as I prepared for my first trip to Russia, I had to borrow travel money from Gene Kimbley. I was not broke, and had avoided bankruptcy, but Silviba was cash poor from overdue receivables.

The Waskesiu fire

This fire was 1½ hours north of Prince Albert and wedged between Highway 2 North on the east side and the Prince Albert National Park boundary at its western edge. At the centre of the fire salvage area, a large muskeg reduced access to some good stands of sawlog spruce up against the park boundary. Carrier sent in a contractor with feller-buncher and grapple skidder on one old road that bisected the proposed logging area. This started out on the highway, and ended in front of the Park boundary.

The Park boundary itself is bare ground. It is easy to locate, since the authorities had cleared a 5 metre-wide strip of brush on the Saskatchewan side all the way along and then fenced it. The Park had removed the trees. The fire had crossed into the Park at one point, but had died out because of lack of fuel.

The contractor logged out the timber from the accessible area. It was his plan to walk his machine across the muskeg to the timber on the other side. This reached to the Park boundary.

However, as the operator started to walk his machine across the muskeg, water flowed onto the tracks. I feared of big machines falling through the muskeg. On previous occasions, operators were severely injured or killed. So I called the operator back to firm ground. I then scouted an alternate route to the timber up against the Park boundary. I walked the machine down the cutline, while the operator straddled some markers on the way. The snow depth was two feet, so the foot print was minimal. The buncher then continued to work and remove the remaining timber. The grapple skidder delivered the wood back out to the main road.

Not long after, a howl went up from the Park staff. Given that it was deep winter and a low point in the Park's usually busy schedule, the staff had nothing better to do but create a significant fuss. Anyway, I ended in court to face charges under the National Parks Act. Mitch Holash attended on my behalf and I got a 'spanking'.

In due course, the Park foresters hired Silviba to survey in a forest fire buffer up against the village boundaries. I had a crew cruising and mapping the timber in the zone of interest. I did a business plan, including cost-revenue projections. In time, the Park logged out the conifer part and left the hardwoods standing. It was a job well done, and lessened fire hazard and risk remarkably. National Parks have this split personality. On one hand they need to perpetuate a wilderness. On the other hand they need to manage values at risk.

As of July 15, 2015, I hear that Waskesiu in Prince Albert National Park has been spared from the current forest fire storms in Saskatchewan.

ECGI (EcoDynamics Consulting Group Inc.) is born

Gene Kimbley, that Métis gentleman from Beauval, and I had mutually explored some options for joining forces for some time. Gene was no longer attached to Montreal Lake Cree Nation or, latterly, Woodland Cree Resources, an Indian logging arm of the Prince Albert Grand Council.

On April 29, we signed a memorandum of understanding, and Silviba assets became ECGI's. Thanks to my dad, I owed no one any money. Although cash poor, I kept the outstanding receivables and one truck. To seal the deal, Geno paid me $1. Jason Nelson was also part of the deal. He brought his connections to the team. In turn, Geno arranged with aboriginal funding agencies to borrow money to buy an Esri[70] [71]GIS. He arranged to pay me $35,000 in monthly payments for the Silviba assets.

The working relationship was rocky at times. Geno did not do well under pressure. At those early meetings with Jason and me, he fired the puck at us when he lost his cool. Once I understood what was behind all this, one day I disarmed him by breaking into laughter. After Jason joined in, no more pucks ever whizzed our way.

My workload doubled immediately. My first assignment was to build an ECGI business plan that Geno could take to bankers, funding agencies and equipment outlets. The project base became more complex as the forest side of the equation shrank in favour of the environmental side. Jason brought soils, vegetation, riparian, wildlife and other features to the table.

Eventually, ECGI received a skimpy line of credit from the Royal Bank. Brenda Abrametz, a genius at helping small companies through the shoals of cash droughts, moved her desk into the ECGI office. She referred to herself as a 'bookkeeper'. She steered Revenue Canada away from causing some serious damage by being on top of each problem as it arose.

Gene brought several Indian forestry initiatives to ECGI. Jason Nelson soon got booked up with projects all around western Canada. On one project, extending soils mapping northward in Saskatchewan, he found new soils associations that had never been recorded before. He loved to dig soil pits. With his trusty spade, he could get down 1 ½ metres or more. By studying the various soil layers exposed, he knew exactly which soil category was which. Thus, forest soil classification became his mantra.

Russian trip no. 1

Out of the blue, I received a phone call from a highly placed forester in Vancouver to consider joining a consulting team tasked with finding ways to improve the Russian forest fire management systems. This was a European Union project, so I renewed my British passport of 1949 and became a citizen of the European Union. Clearly, this helped the general contractor to get the contract. Only EU citizens could take part. Geno drew $200 cash using his Interac card since I was broke. From Saskatoon, Air Canada deposited me at the International Pierre Elliott Trudeau airport in Montreal, but without my 71 kg of luggage, on May 7.

The Aeroflot flight to Moscow was three hours late in leaving. All the passengers and crew sat in a little waiting room. Mums, kids, business people and probable KGB types all waiting for delivery of the aircraft to the gate by Air Canada maintenance staff.

Eventually, the Russian pilot signed the delivery slip provided by an Air Canada maintenance crew member. Air Canada had the contract from Aeroflot to maintain its aircraft and prepare them for flight. Everyone staggered up the stairs to board the aircraft hefting enormous shopping bags. Clinking glass bottles against glass bottles added to the holiday atmospherics created. The overloaded shopping bags squeezed their way along the aisles, the handles hung up in the armrests as passengers continued to their seats. They must have brought aboard half of the Walmart inventory in parcels, containers, boxes and rolls. No sooner had the plane left the ground, the visiting started. Folks greeted other folk to regale one another with the highlights of their Canadian junket. The flight attendants took drink orders for the few that did not have their own bottles. It was all so informal, friendly and classless. The flight was only half-full, so I sprawled across three seats in my attempt to get some sleep.

Going through customs at Sheremetyevo airport in Moscow close to noon on Saturday morning, May 8, had its moments. I became part of a big queue that trickled through passport control on its way to customs. The Aeroflot baggage staff took down all the details of my lost luggage. A contact woman provided a phone number and I, in turn, gave her the name of my destination hotel – Hotel Kozmos. I was released from the confines of the baggage area, as if shot through a cannon, out into a moving, mixing crowd behaving like a school of spawning salmon, all intent on reaching the sidewalk outside. I glimpsed a man holding up a TAESCO[72] sign, and followed him to the car, a left-hand drive Toyota. Russia drives on the right.

On the way in, I had some conversation with the driver, Nikolai, mostly to find out where we were going. The driver's English was primitive, and my Russian acuity was limited; although, I had taken Russian lessons from a Ukrainian woman in Saskatoon. The driver delivered me to the hotel on Prospect Mira, north of Moscow Centre. He would return on Tuesday morning at 8 am to take me to the project office in Pushkino.

The lost luggage blame game continued. Air Canada blamed Aeroflot, and Aeroflot in Moscow searched diligently for this luggage over the next three weeks. An Aeroflot woman in Moscow kept in touch with me regularly. Air Canada continued to snub me. I found out six months later that a theft ring embedded within Air Canada's Toronto luggage system had done the deed. Never an apology from Air Canada, just an instruction to apply to my home insurance for compensation.

The Hotel Kozmos was built in 1980, in time for the Moscow Olympic Games – the games that many countries boycotted. I had the weekend to find shirts, socks and underwear to replace those in my lost luggage. I went down for breakfast next morning and was astounded at the array and complexity of food choices. The arrangement of tables was interspersed with steam tables, griddles, toasters serviced by chefs, all forming a large square within a vast room. Cold cuts of all kinds and breads of many origins rounded out the breakfast fare, including porridge.

After breakfast, my first task was to buy replacement clothing. With my skimpy skill in Cyrillic, I had to prepare by looking up key words in my dictionary. Without the Internet, but with a good map, I found myself in a pedestrian tunnel linking the hotel side of Prospect Mira with the metro station VDNKH side, on the edge of Stalin's technology park. The tunnel contained a myriad of little kiosks, shops and stalls that sold anything and everything. It was crammed with pedestrians travelling in both directions. The lighting was poor. I found underwear on a table in the tunnel manned by a middle-aged woman in black. She chose for me exotic underwear, stepped out into the masses to hold it up in front of me to ensure that it would fit. We blocked all the foot traffic from proceeding further until she was satisfied that I had the right size. I trundled out my rubles and paid her. Next, I went looking for a facecloth, toothbrush and toothpaste. Instead of a facecloth, I made do with a small towel. Russia has no such adornment as a facecloth.

I bought shirts and tee shirts at a recognizable shop further out. The rest of the weekend I practiced reading Cyrillic and visiting the various expositions in Stalin's park, including the Cosmonauts' museum. Since Monday May 10 was a national holiday celebrating the Russian victory in World War II, I had another day off. On that weekend, Matthias Rhein, the project liaison person, visited in order to pass on an updated contract. All rates were cut by 1/3. What could I do but sign the document – I was trapped.

Ω

Nikolai picked me up on Tuesday morning, with the interpreter and other members of the team. We drove up the highway for ¾ hour and arrived at the Russian Aerial Forest Fire Centre. On that first trip to Pushkino, Nikolai switched off the engine during long stops at some intersections. Large black sedans with opaque glass windows drove against the traffic in the median. This narrowed considerably along some sections. Occasionally, Nikolai had one wheel up on the sidewalk to get around stalled vehicles. Halfway to Pushkino, on the left, stood gated communities behind high brick walls. The houses were all two storeys and had the same basic design. The daily taxi ride passed through Tarasovka. Here, the road made an elbow turn to avoid a Russian orthodox church under repair. The dome roof had a new coat of gold paint, but much construction debris lay around the building. Houses alongside the road here were old and built of logs. They had those beautifully carved gable roofs.

The Aerial Forest Fire Management Branch of the Russian Forest Service (*Avialesookhrana*) [73] found in Pushkino was housed in an old building with walls of thick granite. Obviously the previous

346

domain of a long-dead significant someone back in the early 1800s. The front passageway opened into a large room with a flagstone floor. A counter ran partway along. There, a little old babushka provided everyone with a key to the project office on the second floor and had us sign the register. The general contractor based in the UK, Hunting Surveys Ltd had the EU contract. The project manager (retired colonial forester from Edinburgh) and a British forester added the colonial flavor. The general contractor's satellite communications manager (a Polish woman specializing in digital imagery) was responsible along with the Russian counterparts for installing satellites. These downloaded NOAA imagery for forecasting fire weather. They were joined by a Finnish forest entomologist who provided forest health expertise and a German liaison manager who spoke Russian also was a TAESCO consultant. I came up in the rear, the only Canuck (forest fire management geek). The project terms of reference included the creation of improved forest fire initial attack protocols, mainly in southern Siberia. An improved satellite system to receive NOAA [74]satellite imagery became a key tool to improve fire weather forecasting. A key person in the group was the fulltime interpreter, Elena.

The office was a renovated upper floor with new lighting, electrical and windows. It was a pleasant atmosphere. I had two jobs. First, to instruct the group in the mechanics of the Canadian Forest Fire Behaviour Prediction System, together with the Canadian Forest Fire Danger Rating System. Second, externally, to receive information on the present state of the Russian forest fire management situation. The liaison, Matthias Rhein, connected me with a fire science professor at Moscow University. So Elena and I spent a morning visiting him in his office. The professor had written a treatise on the subject, so I had some support in translating important passages later. Elena showed me how to buy metro tickets after one disaster on my first weekend. I had queued up at the ticket wicket at the NKVD metro station. The lady behind the glass babbled at me when I asked for a book of tickets. She grew more frustrated and the queue got much longer – a huge line extended outside. Frustrations rose and she finally took my money and gave me tickets. Elena explained the ticket lady tried to tell me that as a pensioner I travelled free.

Ω

A week later, the team flew from Moscow to Irkutsk in southern Siberia on May 24, just north of Lake Baikal. We travelled all-night and flew through five time zones, to arrive at 9.30 am Irkutsk time. As the Boeing 737 descended, I realised that I was in a remarkably different ecosystem. Steep hills and mountains to the south, lightly clad with trees, marked the boundary with northern Mongolia. The taxiway to the terminal was endlessly long, bumpy and lined on one side with derelict airplanes. They all missed engines, parts of the fuselage, a wing or two as well as cabin doors. It was depressing. The team was taken to a hostel of some kind for a rest prior to a big meeting with key staff from the Institute of Solar and Terrestrial Physics, followed up with a tour. Lunch was soup and funny little dumplings.

The mission was the establishment of a NOAA satellite reception facility tied into enhancing an early forest fire danger warning system for that region.

Ω

The next morning, the team split into two groups. The NOAA satellite installation component stayed back at the Institute, while the fire guys went to *Avialesookhrana* headquarters of the Irkutsk oblast. I received a briefing on all facets of the regional forest fire operations, from initial attack to aerial operations with helicopters and water bombers. In fact, there were two MI 8 helicopters for the entire Siberian forest fire region, and a few fixed wing airtankers. These, to cover an area the size of British Columbia?

After lunch, the team members visited the Forest Management HQ, and met Anotaly Phileminov, the head of forest fire regional organisation. I inspected equipment and handtools, but witnessed sticks used as shovel handles. There was only enough footwear for forty-five men at once. The men had to turn over the footwear to their replacements at the end of each work tour.

After lunch everyone reconvened at the Institute. Matthias recommended that we fill our stomachs with bread before the meeting

348

at 1.30 pm. I found a bread shop and followed his suggestion. Why? The first item on the agenda at this meeting was to bring out the vodka glasses for toasts to everything that moved on Earth. In fact, this was the only item on the agenda. It was fascinating to recognise that after three glasses of vodka, everyone could understand everyone else. I took my turn to make a toast, stood up and 'fluently' displayed my appreciation for the opportunity to tour the Institute and meet Dr. Nickolai Minko and his staff. It was a friendly and warm occasion. Communication centred on body language. This, of course, is universal.

On May 27, the next day, the forest fire members of the team went to *Avialesookhrana* to meet with officials responsible for setting up this TACIS[75] project. This was a candid discussion, and at one point the team was labelled as 'parasites' by an official. Given that a forest ranger and his family were trapped behind a forest fire, perhaps the official showed some anxiety that coloured his judgement.

The next day, the team started back at *Avialesookhrana* armed with a bottle of good vodka, and luckily, Matthias ran into both the Oblast governor and the chief of *Avialesookhrana*. High-level government officials from the Russian Forest Service were also present. Fences were mended, and the chief as well Dr. Nicholai Minko from the Institute and team members were driven to the shores of Lake Baikal for a fish fry and party. I was forewarned by Matthias that wood ticks abounded.

On the beach, the language issue disappeared once more as the vodka flowed. I secretly tipped over my glass at one point to lessen my vodka intake. However, an eagle eye among the Russian unit caught the act. I was yelled at and told to refill my glass. A group of schoolkids from Ecoutian School were out on a day trip. The teacher could speak English, so I had a good visit with the kids. We dropped in at a mink farm on the way home. On its return to Pushkino, the team found that precious computers and satellite equipment had not yet cleared customs. This was to be a thorn in Matthias' side for most of the project.

Once back at our accommodation, we searched for ticks in intimate places. Later I woke up from a snooze to feel a subtle sensation at my hairline. It was a wood tick.

Ω

The social life within the group was haphazard. On one Saturday night, I accompanied the Finnish person to a party in a district close the Canadian embassy. The host was a fellow Finnlander who spoke good English. The party included the owner of one of the largest Moscow radio stations. His fiancée was a Moscow university forestry graduate specializing in translation, Elena Kopylova. After the party, Elena and hubby-to-be dropped me and the Finnlander back at the Hotel Kozmos. Elena stayed in touch and exchanged emails. This was to be an important contact on my last tour of duty in 2000.

I returned to Canada on June 8.

Russian trip no. 2

On October 1, I started my second tour to Russia. The view from my perch in the hotel was a gift. I could look down at the hustle and bustle below. Prospect Mira was eternally busy. A cross street containing a cluster of bus stops for all routes passed through an open vista of grass interspersed with exhibition buildings in the technology park. I watched the start of rush hour around 7.30 am each morning as passengers streamed from the bus stops over to the metro. All the way along, this path was lined on either side with vendors of cigarettes, perfumes, clothing, artwork, figurines, newspapers – you name it. The vendors just stood with their goods; there were no tables or booths. At about 9 am, the flow of commuters tapered off and the vendors dispersed.

It was widely believed the rooms in the hotel were bugged. Ominous men strolled through the foyer, and police with AK 47s stood at all major intersections. However, this was President Boris Yeltsin's reign; the country had found free enterprise. Once more, I became part of the daily project regime. I briefed Matthias on my trip to Winnipeg in the summer to visit CIFFC[76]. There, I had related my Russian investigation and received feedback on this agency's experience in dealing with their Russian colleagues. CIFFC's main task was to coordinate aircraft and ground crew positioning depending on the forest fire danger rating in the various provinces, and even Australia. The resources moved around to match the intensity of fire danger anticipated. A crew from the Ontario forest fire authority may end up fighting fires north of Prince Albert, and vice versa.

During this tour, I met officials from both the Canadian embassy in Moscow and USAID. This latter agency was working with *Avialesookhrana* in Novosibursk. I received information that was mainly to do with the provision of forest fire equipment.

Ω

Chapter 28:

EcoDynamics (Silviba), 2000

I now worked for two outfits. I was Geno's coach, mentor, slave and whipping boy. But Geno delivered. In May, ECGI bought the international standard for GIS software, ARC INFO from Esri Canada, and new computers. Once more, ECGI could compete for both government and Weyerhaeuser projects, among others. Joachim Boehm, an immigrant forestry graduate from Germany entered the ECGI world via my connection with Aallcann Wood Suppliers Inc. Bob Romanchuk, Aallcann's principal, used ECGI in to prepare Aallcann cutplans. Aallcann became a sleeping partner through cash contributions. It was a synergistic relationship. ECGI needed a GIS operator. Aallcann's forester, Joachim Boehm learned to use GIS to prepare the applications for fencepost permits, including reconnaissance and map work. ECGI could now handle all levels of mapping, from soils to riparian. Also, all maps needed for both forest and environmental planning. Jason Nelson still commuted from Saskatoon, but he would move to Prince Albert. Roger Bonneau, a forestry graduate from the U of Alberta was in his second year as a junior forester. He lived in Debden, a fair commute. Also, early in the year, Gordy McRae rejoined, this time as an ECGI employee. Bob assigned Brenda Abrametz to do the ECGI accounting.

Brenda is a unique individual. I got to know her when she served as bookkeeper for Barry DeVocht, Provincial Forest Products Ltd after Barry had moved into the office beside Silviba. She is truly a gift in that she keeps companies afloat who suffer from *cashflowum interruptus*. One manifestation of this disease is a lack of immunity to

Revenuum Canadensis. Without Brenda, PFP would not have lasted as long as it did. And latterly, she played the same role with ECGI. Brenda is truly unique in her skill set. She grew up in Arborfield, SK. – *Brenda, huge hugs are sent your way.*

Brenda married Peter V. Abrametz, lawyer in Prince Albert during her PFP sojourn. He, in turn, creates large footprints that at times pick up a bit of mud.

Ω

Geno brought in work from both Alberta and Swampy Cree nations on Saskatchewan east side. He had mentored the start of Delta Management Co., formed to become the logging arm of the Cumberland House community[77], a non-member of the Prince Albert Grand Council. He was also on the board of directors. This community was a hybrid mix of Métis and treaties. It is the site of a Hudson Bay post, the fur trade gateway to all points west. We helped a First Nation to set up a sawmill, buy logging equipment, take out cutting permits, plant trees and otherwise join forest-based economies. Geno found logging contractors, including Jackie Ross.

I put together training plans and did the paperwork needed for grant applications to funding agencies for equipment and training. It was not an easy row to hoe. Stresses and strains within the community contributed by friction between families, and the ambiguity of Treaty vs Métis, complicated the discussion at the board level. However, timber was logged, markets were found, and cashflow ebbed and flowed. Geno was cheque signer, so he picked up added stress at times.

Geno also picked up work with an Indian band in Alberta, and later with the Stoney band west of Calgary. Meanwhile, I had to service my own clients as well. On February 25, I returned to Moscow for trip no. 3.

Russian trip no. 3
I arrived in Moscow minus a small suitcase. The weather was wet and snowy. I had the routine down pat. A big construction project was underway on Prospect Mira. I saw the activity from my upper

storey room. The same folks lined the pathways between the bus stops and the metro. Their wares had now expanded to include women's fashions as well as the usual trays of cigarettes. An overhead roadway was under construction. This would lessen the congestion at the road junction across from the metro. The grade started about 300 metres on the south side of this Prospect Mira junction. It was an excellent design choice since it climbs a slight grade to that point. The engineers just carried it along over the junction then tunneled under the next junction. I saw the pilings driven into the ground; massive concrete columns built using cages of rebar, and endless convoys of cement trucks. The concrete stringers arrived, prebuilt, on large tractor- trailers. The crane operators then strung these across each set of pillars to form the platform for the steelwork to follow. This all went on, day and night. Squads of workers in hardhats and yellow jackets scurried from one set of footings to another as they prepared the bed for the next concrete column. They had placed the construction office immediately across from my hotel window. It stood right beside the inventory of rebar and welding station. The welding flashes danced and flickered all night and lit up the room. During this construction, the southbound lanes of Prospect Mira moved to the west. The east front of Stalin's technology park lost its boulevard trees and sidewalk.

The morning commute always included stalled Ladas at the roadside somewhere en route. I was rather shaken up to see an ambulance at the edge of the road with the hood up. On another occasion, it was a fire truck. My assignment included preparing a paper on emergency response improvements. I studied the Russian Forest Fire plan for 1999-2005. I needed to incorporate recommendations for both improving emergency response and initial attack response protocols. The team shuffled around. Therefore, the overflow used a desk in Ludmila's office next door. Ludmila was the Russian connecter to *Avialesookhrana*. One morning, I came into work in her office and found that someone had stuffed 3M tape into my CD drive overnight.

Work also went on back at the hotel, which served as the team base. Evening discussions erupted. Often, these flowed out to the various restaurants as meetings continued over a meal. Matthias invited me for supper with his family one night. Masha, his wife, and little Amun entertained me generously. Masha's English was excellent. She translated several articles that I needed for my work. Not soon after, Matthias resigned from the team as Masha was not happy in Moscow, even though she was a Russian. She did not want to bring up children there.

Moscow contained 14 million souls in 2000. When comrade Brezhnev took over from Khrushchev in 1966, he sped up the construction of apartment complexes throughout the city. These were constructed of concrete and are clustered as four towers with a grass play area for kids in the centre. Of course, within the older city environs, many old buildings were converted to apartments. From the Hotel Kozmos, these clusters coated the newer parts of the city in every direction. Clusters with a minority of children tended to replace the grassy courtyard with a sea of asphalt for a parking lot. Car alarms went off often, and this sound bouncing around the courtyards and penetrating bedrooms at 2 in the morning was an aggravation. I loved to see the kids and their mums walking to school. The little girls wore dresses and their heads were adorned with white – always white – bows and fluffy accoutrements.

Goertz Schuerholz, a connection from TAESCO, joined the team on February 22. He, Antti Puttu, Robin Cutler and Martin, the Scottish consulting forester and I tried out several restaurants over the next two weeks in our search for a good supper. As a Canuck, I was dismayed at the restaurant behaviour displayed by the Brits. It seemed the British Empire had not completely faded away, even at a small restaurant in the Ostavinsky district.

Beata, the HTS[78] GIS technical guru, now came to stay. Beata's family had escaped from Poland to Great Britain at the start of World War 11. She and I had several good discussions over the next while. I began to suspect the HTS mission was to install the satellite receivers; everything else, including my contribution was peripheral at the best. 'Window dressing'?

I completed the 'Emergency Response' paper, as well as the 'Prescribed Fire Protocols' paper before I returned to Saskatoon on March 28.

Indian Special Claims: 1906 timber surrender – Sturgeon Lake First Nation

Chief Earl Ermine had written to Silviba in April 1997. The band, one hour north of Prince Albert on the south side of Sturgeon Lake, needed a forensic forester to explore the theft of commercial sawlog material from the Sturgeon Lake reserve in 1906. Under the Northwest Territories Indian timber regulations, 1887, no one could remove logs from a reserve without a permit from Indian Affairs and permission from Chief and council. Indian Affairs collected the Crown timber dues payable under this Act and held them in trust for the band.

On the band side, Harold Kingfisher, an elder and a former chief, and Jack Long, the diligent and overworked band administrator, became the Sturgeon Lake representatives. Alexander Dietz, an independent Indian researcher with connections to Indian Affairs archives in Ottawa contributed his skills.

I had to stand the timber back up that was on reserve in 1906. To do this, I accessed the Department of Interior timber agent's records from Saskatchewan Archives Board and Indian Affairs files. Jason Nelson did the forest soils mapping on reserve. The Federation of Saskatchewan Indian Nations (FSIN) research staff contributed research assistance as well. The soil classification was matched with the forest ecotype occurring in adjacent standing timber. A timber cruiser's map from 1886 helped greatly. The area involved was then matched with the Saskatchewan timber volumes within the C-50 forest inventory zone. The archives contained all the correspondence and files for the successive timber agents within the Prince Albert district, Department of Interior. In April 2,000, Chief Wesley Daniels and band agreed to a settlement of $4.4 million. The complete story is covered in Part Three.

$$\Omega$$

On August 16, Geno and I travelled to Cumberland House to meet with its forestry board. It was an exhaustive meeting, and an even more exhausting trip home. Geno drove, all the while chatting his head off – to keep himself awake, one suspects. Anyway, I arrived home at 0330 hours. I had to be in the office early the next morning to write up a critical invoice. At 0730 hours, I was just about to turn on my computer at the office, when I experienced an inexplicable dull ache in my chest. I wasted no time dialing 911 before heading back downstairs to wait on the street for the ambulance. I held on to a tree and vomited. Shortly, an ambulance and fire truck showed up. The ambulance crew drove me up to the Victoria hospital after they tried to figure out the route from downtown Prince Albert. The debate on the way was energetic. My skin was wet with sweat, so they could not get a needle into my arm. Later, I found out that my heart attack was serious.

I stayed in the ICU for five days because of atrial fibrillation. During this time, a cardiac rehabilitation therapist dropped off tapes and a TV for me. Unknown to me, Eddie Kwiatkowski's wife, Brenda, a nurse at the hospital, had made sure that I got a room to myself after leaving the ICU. There, I had regular visits from my key staff to guide them as they worked on the various projects. Since these visits fatigued me considerably, I booked off frequently. The nurses posted a 'no visitors' sign on the door. Also, I exercised every day by walking around the fifth floor.

On September 22, after I was checked out by the heart rehab staff, I started heart rehab at the gym just a block from my office here in Prince Albert. Of course, as a new recruit I wore the orange tee shirt provided. And attended the compulsory nutrition class afterwards. The sessions were on Mondays, Wednesdays and Fridays, and started at 0830 hours. We undertook a suite of exercises including the rowing machine and stationary bike. After nine months of this, I graduated to a green tee shirt and did not have to show until 0900 hours.

All sorts of folk attended. As we walked around the gym doing our daily routine, we got to know each other's medical circumstances. We knew what medications each of us was on and what doctor said what about our individual heart conditions. Some wives exercised with their husbands. In a subtle way, it became a group healing arrangement. Also, the doctors took turns attending each session supporting the heart rehab nurse and physiotherapist.

The staff kept telling me to slow down. Twice on the bike I collapsed and ended in emergency. I learned what 'syncope' is all about. The trigger is low blood pressure caused by dehydration and low electrolytes. I ended up in Jasmine's lap (the supervising nurse) both times before I passed out.

I cannot express my feelings on heart rehab loudly enough. It is essential to one's regaining heart health and an improved self-disciplined quality of life. I was totally blessed by that program at Victoria hospital in Prince Albert. Indeed, knowledge is power. Very few attendees passed away in due course. My special reward was coaching new folk and giving encouragement as they gradually worked their way back to optimum health. Thank you, God, for the opportunity to join this heart rehab program at Victoria hospital. It provided both Hope and Healing.

Four years later, at St. Paul's Hospital in Vancouver, I found out that I had lost 35 percent of my heart muscle. And so on.

Stoney tribes, Morley, Alberta

Geno picked up the Stoney tribes as an ECGI client. Their reserves straddled the Bow River Valley between Cochrane and Canmore in Alberta, along the Trans Canada. Good lodgepole pine and spruce coated the hills. Access was no problem. In the lumber boom of the early 1990s, logs were scarce all over. Slocan Forest Products, with several mills in British Columbia, was desperate for timber. This company rounded up sixty rail carloads of spruce logs monthly from private lands in Saskatchewan at the time. Somebody in the Stoney Tribes put two and two together, and then found a logger. Shortly after that Indian logs joined the exodus to British Columbia. At least one savvy Indian became rich.

Eventually, Indian Affairs tried to heal the obvious logging scars on both sides of the Bow River Valley by aerial grass seeding. Chief and council wanted their trees back. So they asked Geno how to do this, and Geno took me up for a helicopter ride on an inspection of a raped and pillaged forest landscape. Cutblock boundaries were shaped to attract windfall at the timber edge, not to minimise it. Logging was 'dirty', with trees abandoned where they were felled, and the landings left with good logs left behind. No forest would renew itself on ground sealed with grass. I got the helicopter to land in several spots and we performed ground inspections in a few critical areas. No new tree seedlings anywhere.

The Stoney tribes were rich. Two natural gas pipelines passed through their reserves in the valley. The community collected a royalty from each cubic metre of gas transported across the reserves. Each young person received a cheque for $40,000 on their sixteenth birthday. Car dealers in Calgary trolled for these reserve kids with windfall cheques in their pockets. Several youngsters wiped out as they drove home with their new vehicles.

Gene and I returned home to design, then cost out, a forest renewal plan to replace the logged trees with healthy spruce seedlings. In lodgepole pine ecotypes the ground treatment involved site preparation. Screefing equipment would break down the slash and expose mineral soil. This in turn would promote natural seeding of lodgepole pine. The forest nursery at Bragg Creek would grow the planting stock needed to renew the forest lands logged. It was all doable and would provide months of work for young band members and not-so-young equipment operators. Spread the wealth instead of stealing it.

Was this ever done? I doubt it.

Weyerhaeuser

ECGI became embedded within the Weyerhaeuser Timberlands function. GIS operations, data loading work and trials with map updating using orthophotos[79] became part of the company's revenue stream. ECGI (Silviba) developed a Q-&-A analysis for evaluating the sampling accuracy of forest cover boundaries (ecotypes). Other features displayed on the new series of contracted forest inventory maps needed testing in the field. It was up to ECGI to recommend a 'pass or fail' before the payment of these forest inventory invoices.

Map updating with orthophotos used to delineate cutover blocks followed. Add in beaver floods – in fact, any shrinkage from the total forest inventory. The provincial government used the information developed to confirm that Weyerhaeuser's forest management met its prime objective – to maintain a sustainable forest ecosystem. Remember, you have to regrow what you mow down.

ECGI also received a contract for the annual maintenance and remeasurements of the WESBOGY[8081] research plots. These were established on cutover land with the purpose of measuring both growth and change in the tree species present over time. Every plot boundary was demarcated with paint and every tree in each plot tagged with a number from 1 to 300 or so at a point 1.3 metres above average ground level to obtain a dbh.[82]Unlike humans, whose chest measurements are independent of the height above ground. Also, every tree height was measured annually as well. This job took a full week with a three-person crew in the spring prior to green up. Some sample trees were planted spruce, others were natural ingrowth, and by far the numerous were the aspen. These latter trees came from rhyzomes endemic to all boreal forest cover. Remove the spruce canopy, let in the sunlight to warm the soil and you will observe a sea of young saplings emerging all over.

A long pole helped with the measurement of tree heights. Once the treetop grew beyond reach, you had to revert to a height measuring instrument with laser technology. These are accurate to a decimal of a centimeter, but my shaky hand could not hold one still. Miro was an excellent tree height measurer.

Since the crew always carried the previous year's data with them, occasional errors in measurement appeared from year to year. However, penalties imposed included redoing a whole day's work if too many errors showed up in the check measurements. Also, an ingrowth tree was not a tree until it achieved a height of at least 1.3 m. Thus, once in a while a crew would miss a new candidate. It is fascinating to observe the annual changes within new forests. This is an intriguing example of silviculture research. A spruce from natural seeding grows at three times the rate of a planted spruce seedling. The only problem is time. The ingrowth is dependent on the nature of the seedbed. Also with the extent of new seed crops from bordering mums and dads.

FEC – Forest Ecosystem Classification

Adventiously, my earlier all-day meeting with Steve Smith; the Weyerhaeuser vp for Saskatchewan Timberlands showed some results. I had tried to make a case for Weyerhaeuser's adoption of ecosystem classification within its forest inventory database. Now, the Company was well on its way to incorporate all the ecosystem features necessary for more accurate modelling of forest growth and yield. These tools for measuring sustainability now became essential.

Mike McLaughlan, the forest ecologist within the Saskatchewan Forest Service, introduced the FEC project.[83] This incorporated forest sampling of all Crown forest ecosites in Saskatchewan. Surveyors did much of this under contract over the next several years. As the website suggests, the team undertaking this work contributed various skills in developing an FEC for Saskatchewan. A botanist, a biologist, a forester, a forest soils specialist made up the team.

ECGI contributed a significant portion of the sampling protocols needed over the next several years (http://www.environment.gov.sk.*ca*/ forests). The FEC program used a stratified random approach to sample plots within each stratum.

And Jason Nelson with his unique talents led his team. He recruited his key specialists from all over, including Alberta, Saskatoon and rural Saskatchewan. These folk could exist comfortably in all but the wettest conditions. One evening way up Highway 102 just south of the turn off to Southend, we camped in a clearing overlooking Brabant Lake. Each of us had our own little tent. The ground was rocky. Just before supper, the wind became a gale and each of us had a difficult job of keeping the tent tied down. In my case, I lay all night against the tent wall to prevent the tent from sailing away with me in it. What a night!

In 2000, at the start of ECGI , cash was tight. Geno, in his new role of Company president and general manager, worried over cashflow hassles. The October payroll that year was $15,000. My little stipend of $1,000 did not even show up on the landscape. Jason was hugely frustrated since he had just assembled his crew from all over.

It is fair to say that Jason's drive to examine forest ecosystems and dig soil pits all the way down to Australia added to the general stress during that early ECGI startup phase. Jason is a perfectionist. Luckily, Bob Romanchuk's infusion of $12,000, the first of several infusions, allowed Jason's crew to go to work.

The pre-occupation with cashflow, skilled staff and invoiceable projects was the prelude to the commencement of Prince Albert Model Forest Phase Three to follow.

Ω

Chapter 29:

Soup to the end in 2003

Final Russian trip no. 4

On March 18, 2001, I set out for the last time to Moscow. This trip was much more 'ad hoc' however. Remember I was a sub-subcontractor, and my employer, TAESCO, a German consulting company, was a subcontractor to Hunting Technical Services, the British general contractor.

Officially, TAESCO's part had finished by 2000. My last trip was an afterthought. Without this final contribution, the project was incomplete. HTS would not get paid by the European Commission. The delay, of course, was due to my heart attack in 2000.

This time I travelled Air Canada from Toronto to Frankfurt, then boarded a Lufthansa Boeing 737 to Moscow. This time, I arrived with all my luggage. Nikolai picked me up at the Sheremetzovo terminal as usual, but I missed that wonderful experience gained at passport control on my previous visit, for I had a facial crease that showed up as a diagonal scar across my forehead. It was just above my left eye and, seemed to show up when I was stressed.

Anyway, this neatly uniformed and beautifully packaged young passport woman stared hard at me as I stood in front of her glassed-in counter for some intense moments. Even to this day, I can recall that warm feeling the young women imparted as she gazed right into my eyes. In return, I gave her a huge smile – the spell broke the contact and she waved me on with a mischievous smile in return. It took some time for me to understand the facial crease did not show up on the passport photo.

I arrived at the hotel to find that no one had booked me a room. Robin Cutler met me in the foyer and helped me to get a room for the week. After that I was on my own since everyone was going home. The Kozmos was booked up after that. I had to pay ahead for the week with my American dollars. The hotel desk folk only took brand new US$ bills – nothing creased or used. To get this money, I took my American Express travelers cheques to the exchange facility on the mezzanine floor. Here, I parted with my cheques in return for brand new American dollar bills. In those days it was cash only. Back at the desk, I trundled out a sheaf of bills, which someone then inspected with ultraviolet light before I received a receipt for a week's accommodation.

That week was also the final week for the team's interpreter. Her normal position was in the hotel's business support centre. Here, I had access to the hotel's Internet. I contacted Elena Kopylova and asked her to meet me one evening in the hotel coffee shop.

Elena Kopylova, and a forestry translator student named Tatiana went to lunch with me the following Saturday. The restaurant was on Prospect Mira some distance south of the hotel. It was a great experience to eat with people who could translate the menu. I had a delicious bowl of real borscht with a lamb stew. Elena Kopylova had arranged with someone in the Russian Forest Service to provide me with a room at the Forest Service hostel at Pushkino for the next two weeks. After that, Forest Service staff from all over Russia filled the hostel while they attended various courses. Tatiana's parents would let me stay for the final two weeks at an empty apartment in Ivanteyevka[84], a half hour bus ride east of Pushkino.

So, by Saturday evening, Tatiana's parents had driven me to the Forest Service hostel in Pushkino where I settled in comfortably. The walk to my office in the *Avialesookhrana* building was about half an hour. Since I had a fridge in my digs, I stocked up with breakfast and lunch supplies. My daily walk to the office took me past the speed bump on the main road into Pushkino from the east. Here, I watched as the daily commute scattered Lada parts indiscriminately along the roadway.

For the remaining portion of my contract time, I had the office to myself. I greeted the babushkas behind the counter at the entrance and signed in daily. I had the freedom to visit the *Avialsookhrana* staff, especially the aerial fire staff. Luckily, the boss spoke English, as did one other key staff person. They were young men. Before returning to Canada, I asked them if they would be open to an invitation to Canada to view forest fire facilities and meet with key managers. They were enthusiastic at the idea.

At the end of my stay at the hostel, Tatiana's dad picked me up and drove me to the empty apartment in Ivanteyevka. Tatiana took me up to the apartment on the top (12th) floor; I gave her the rent money for two weeks. She asked that I not open the front door or answer the phone for any reason. She told me she would return the next morning to show me the bus route to Pushkino.

The next morning, she met me outside the building. We walked up through a series of lanes that cut through a very old residential district. These ended on the main drag. We then caught the bus for Pushkino. We wedged past the door into the back of the bus. There, we were confronted by a large spare tire leaning against a post. The ticket collector waded through the packed bus to collect my ten rubles. A half hour later, Tatiana had me get off at the street just up from *Avialesookhrana*. She remained on the bus and continued with her journey.

It was truly precious to walk to the bus each morning through that lane. Houses built of logs and framed with beautifully crafted frescoes lined both sides of this dirt road. Plumb in the middle of each intersection stood an old-fashioned water hand pump. These were always in use as the babushkas packed water back to their homes.

I discovered that I could catch a bus to Moscow from Pushkino. These were 15-passenger buses. They ran regularly and were always full. All routes cost ten rubles (25 cents) one-way. I got off in Moscow at the same bus stops I had noted from my hotel room. The overhead roadway across from the Kozmos was complete, as was the tunnel under that congested intersection one block north. From there, I did my 'old man' thing and shuffled through that free turnstile used by pensioners at the metro. All under the watchful eye of the resident babushka in her grey uniform with red beret. It always helped to pack along the quintessential shopping bag. Mine was from Toyota. It exuded advertising on all fronts – in Cyrillic of course.

Returning to Ivanteyevka after work, I bought food supplies at a modernish shop found on the main drag. As I entered this large space, I noted that a ring of counters surrounded me. Young women in smart smocks served the goods needed from their various workstations. And they all dissolved into giggles as I tried to speak Russian. It was an agreeable experience on every occasion. For many items, I just pointed. They still giggled. It sure beat the sterile happenings within the Kozmos Hotel.

The apartment had a TV. Raisa Gorbechev (Gorbochova) had died in 1999. Mikhail Gorbechev was now a widower. He looked so forlorn during his various TV interviews. During my two week occupancy on the 12th floor, I had an excellent view of the countryside to the east. About thirty km away, an airfield was busy with aircraft continually taking off and landing. My apartment building was the last one on a new road that ended abruptly. Beyond that, it was all small plots and little shacks. I moved in just as someone began to erect a new apartment building just to the east of my balcony.

It was enthralling to view progress each day. They moved in a sky crane first. Trucks started to haul in floors, walls, stairwells, doors and windows. A woman crane operator climbed up the enclosed ladder each morning to start her job of hauling up the bits and pieces. About 1.30 pm she came down for a pee break. After they laid each floor, they erected all the walls for that floor. Welders welded the corner brackets set into the concrete sections together, followed up with the placement of corridors and apartment walls. An elevator shaft also emerged. The crew laid down a floor a day. The next day, they laid in the ceiling blocks and another set of walls went up. Four, one-bedroom apartments for each floor. This all made the Vancouver construction model appear excessively expensive. When I left Russia for the last time, Russian folks were only just beginning to pay rent.

Elena, the business centre translator arranged with her dad to pick me up from the apartment building on my final day, and get me to my Lufthansa flight for Frankfurt, then Canada.

Before I left, Elena Kopylova invited me to a boat trip in Lake Baikal in May. Regretfully, I declined. Too much to do back home, and who would pay the travel claim?

Despite several efforts back in Canada to follow up with an official invitation, I was never able to swing a visit by *Avialesookhrana*. Eventually, I heard that a large Russian delegation visited Montreal. Elena Kopylova came with them. She emailed me to organise a 'meet', but I could not get away.

Indian forestry initiatives continued

I took Barry DeVocht up to Deschambault in an attempt to get his Provincial Forest Products new sawmill and waste burner moved to the Deschambault junction. It would have suited the local log profiles and would have become the focus for the new Peter Ballantyne Cree Nation Forest Resource Management Agreement talked about for so long. Carrier Lumber had completed its due diligence in the same community previously. Perhaps, those folks controlling the Treaty Land Entitlement investments back on the PBCN urban reserve in Prince Albert looked elsewhere for opportunities. An earlier attempt to create a joint venture between Ainsworth Lumber Co. of Vancouver and the band had also failed. That potential relationship between PBCN and Carrier Lumber in Prince Albert contained even greater benefits. In the late 90s, the gleam went out of the lumber market, so Deschambault timber became too expensive. However, Tolko Industries at The Pas provided training and support to the PBCN emerging forestry arm, Mee-Toos.

This arrangement did come with baggage. Logs were exported to Manitoba through Flin Flon, then down Highway 10 to The Pas[85]. More could have happened, but the Mee-Toos manager had a reluctance to work with Geno, who had tangled with him when he logged for Delta at Cumberland House. A disconnect between PBCN head office in Prince Albert and the human resources within the Deschambault community held up progress with band forestry initiatives. These were there all right because of my training over the last ten years. Jack Custer had seen to that. He had also equipped a Deschambault forestry office complete with forestry maps and forest management plans, and Mary, the secretary. However, PBCN head office was really more comfortable operating motels and owning a casino. By the way, choosing a sawmill site at the Deschambault junction removed thirty km of extra travel. This also removed the need for beefing up the power distribution system that extra distance.

Ω

Another initiative was emerging within several bands closer to Prince Albert. The wood supply was to come from the Island forests – Nisbet, Fort a la Corne and Canwood forests. These were remnants of Crown forest lands in place after settlement just north of Prince Albert. They stick out like a sore thumb on Google Earth imagery. Ron Burns, a band member whose family originated on the James Smith reserve on the south side of the North Saskatchewan River served as coordinator and the focal point for the seven bands showing interest. He set up an office close to Geno's. Funding for all the efforts expended came from sources accessed by Eddie Head, a former James Smith chief, and director for the First Nations economic development fund.

My job through ECGI was to identify economic opportunities, set up a training scheme, identify forest resources available, prepare forest management plans, and then build a business plan. Geno and Ron Burns massaged the chiefs and councilors involved to obtain consensus on the approach to the provincial government. I drafted an agreement for examination by Saskatchewan Forestry Branch and by all involved bands. This designated responsibilities and a financial plan. Rome was not built in a day. It took the better part of the next three years before something was to happen on the ground.

Ron Burns was an essential part of this Forest Management Agreement evolution. His coordination role assisted Geno's efforts to keep the project development moving ahead. Without the ECGI infrastructure Ron would have not have achieved any of the proposed objectives: jobs, cashflow and community business enterprises. To organize a sustainable wood supply from provincial Crown forest lands within a federal Indian initiative was cumbersome. This went on at two levels: with the chiefs and again with the technicians. Geno had to get the technicians for each band into a room to discuss solutions to technical issues. Then these people would return to their respective chiefs and councils with the technical recommendations. Ron Burns then organized a further meeting with the chiefs to ratify the decisions made at the technical level. Eventually, an Island Forests management company did form, and Saskatchewan Environment/Forest Service issued a five-year term Timber Sale Licence with a renewal clause.

Jason Nelson, ECGI's forest ecologist and environmental specialist,[86] developed an integrated resource plan in support of this enterprise. I have to state that I have never seen such a more thoroughly researched result before or since in my career. Jason sought animal biologists, avian specialists, and botanists. He added his own expertise with soils and vegetation. ECGI's foresters and forestry assistants supplied by the bands updated the forest inventory data and produced a five-year management plan. This highlighted the socio-economic component including tourism as well as jobs. Indian Affairs helped pay for this work through grants.

Later, ECGI completed further environmental assessment studies for De Beers Diamond Company in Fort a la Corne and for James Smith reserve. This First Nation explored an opportunity to dam the South Saskatchewan River to create power for supply to the SaskPower grid.
(www.environment.gov.sk.ca/.../adxgetmedia.aspx?).

ECGI relied on the GIS work of Heather Patterson. She was a key member of the ECGI team. Heather became an essential element in ECGI's suite of skills and expertise. I remember when she came on board after her training at SIAST in Prince Albert.[87] My major piece of advice to her was that a map has to answer more questions than it creates.

Wapawekka Lumber
In 1999, Wapewekka Lumber Ltd, a partnership between Weyerhaeuser and Woodland Cree Nations in northern Saskatchewan utilised northern timber reserves. Geno was on the board of directors since he had represented Montreal Lake in forestry matters. This

sawmill, set up with curve sawing technology to maximize the recovery of lumber from small trees never ran efficiently. Coincidentally, the mill also made wood chips for Weyerhaeuser's pulp mill next door. From the band side it was never a success either. Weyerhaeuser struggled with lumber recovery continuously. In 2009 the entire mill was sold to an outfit in Alberta.

Geno never talked about his involvement with this enterprise. I feel that Weyerhaeuser set this all up to prevent other companies from moving into Saskatchewan and capturing wood supply such as the potential PBCN FMA at Deschambault. By selling products to itself, Weyerhaeuser made sure that Wapewekka lumber was sold at cost to its own lumber outlets. This meant no profits for the bands. All rather sordid behaviour from a corporation with such a high horizon.

Agency Chiefs Tribal Council

Agency Chiefs Tribal Council, wedged in by Prince Albert Grand Council to the east, and Meadow Lake Tribal Council to the west, started to build modular homes and create a logging arm. Their leaders employed managers who did not seek the services of Silviba, or ECGI. They had logging contracts with both Carrier and Weyerhaeuser, and between the usual forestry ups and downs, are still busy today. At least somebody has their act together.

Prince Albert Model Forest retools

Early in 2002 ECGI received a contract to retool Prince Albert Model Forest. Thus Phase Three emerged. Some partners, including Montreal Lake Cree Nation faded away. Federal funding shrunk and the heady days of showing off Canada's effective sustainable forest management programming to a critical world slid to the back burner. Because forest management in Canada is the responsibility of each forestry province, the federal government restricts its lip service to Indian reserves, Canadian armed forces training grounds and much diminished forest research.

ECGI took on a management contract and reported to a board of directors. Geno became the manager. Irene Roy, the Prince Albert Model Forest secretary moved into an office beside Geno. Weyerhaeuser, also, wanted a reduced role. Too much staff time with not much payback. I monitored ECGI costs closely and found that over a year, the Company lost money. Just too much baggage to handle with no invoiceable results. Bureaucracies are hard to change. The Canadian Forest Service continued its involvement undiminished.

It was my job to draft the Prince Albert Model Forest Phase Three birthing document in support of continuing the Prince Albert Model Forest program born in 1992. At times I slept at the office to meet deadlines for draft reviews by the board. It was a drawn out process lasting all of six months. There were so many players. The Canadian Forest Service, both in Edmonton and Ottawa had to ensure that all wording was compatible with legislation that supported the Canadian Model Forest Program. The Saskatchewan representatives from Environment, usually dormant, woke up and used the Model Forest as a vehicle for some of their beefs, battles and belligerence. The Federal terms of reference were inscrutable, often vague and capable of misinterpretation.

Rather than add meat to the soup, Saskatchewan Forestry Branch spent its time at various work meetings criticizing my skills with Microsoft Word. A few Forestry Branch staff despised contractors, and this was a God-given opportunity to exhibit their venom. Their basic difficulty was the inability to provide anything useful for inclusion in the Prince Albert Model Forest Phase Three draft documents.

Though, in fairness to all concerned, there is usually more than one side to a story.

Ω

Geno's Model Forest duties extended into participation in International Model Forest programming. He made trips to Finland, Germany and Russia, as well as somewhere off the Peruvian coast. He took Patricia, his wife, with him at times. After Pat returned from Russia's eastern region, she regaled us with her experiences. The most noteworthy being the accommodation in Khabarovsk. There was no washroom, just a hole in the floor.

By November, Geno was again talking of crew layoffs because of lack of cashflow.

Dakota Whitecap reserve fire claim

In November 2002, David Knoll, the Sturgeon timber claim lawyer asked me to examine fire damage to reserve infrastructure on the Dakota Whitecap reserve south of Saskatoon. It is found on the western edge of the Dundurn military training grounds. In October, white phosphorous shells from a tank exercise set the vegetation on fire. A wind then blew it on to the bordering reserve, melting house siding and burning down forest land and hay meadows. I did the damage assessment, calculated the cost of repairs and produced a costed forest regeneration plan.

Ω

Moose Lake Indians, Manitoba

ECGI, in cooperation with Manitoba Department of Resources, Tolko Industries at The Pas and the Moose Lake Indians, developed a forest management plan covering Moose Lake's traditional lands. Geno and I scouted the area and the potential forest products suitable for the economics available. We blended this in with a need for poplar by a small milling and manufacturing company just north of the U.S border at Piney, Manitoba. Then we visited this company to settle the specific log size and grade needed. It had a good market south of the border for cabinetry.

On my return to Prince Albert, I liaised with Manitoba Natural Resources and Tolko. Geno sent out foresters to sample timber stands and check out the Manitoba Resources timber inventory. I then put a business plans together and costed out a small sawmill for placement on reserve. Since the band already logged for Tolko, most of the poplar would come as a byproduct since Tolko did not process poplar.

The band travelled to Prince Albert to review the plan. They then arranged for a meet with funding agencies in Winnipeg shortly. I did the cashflow forecast in preparation for this trip. The meeting did occur, but the stumbling block was an unsettled claim for forest lands within the Tolko lease. So, yet another project with excellent potential died in birth. Heartbreak.

Ω

Agroforestry on private lands

I took on the agroforestry challenge. The existing federal forest resource agreement with Saskatchewan included funding to promote afforestation on lands unsuitable for crops along the northern grain belt. An opportunity to grow wood fibre for use with electrical cogeneration emerged. A provincial committee formed to coordinate agroforestry on private land met in Tisdale early in 2003. Fortunately, the agricologist from the provincial Department of Agriculture in Tisdale, as well as a Canadian Forest Service technical officer in Prince Albert, both committed themselves wholeheartedly. Their mission? Assisting landowners to develop a procedure to achieve a standing crop of fast growing poplar on sites picked for such a use.

My job was to help the landowner to fill out the forms for funding, design an approach for each opportunity. I then helped with implementing the growth strategy approved. Farmers came out of the stubble to get involved. Forest regeneration called for planting hybrid poplar cuttings through sheets of black plastic strips up to a half mile long. A tricky exercise, especially when the wind is trying to blow the plastic into the next field. Since I lived in Birch Hills, and that community is noted for its adventurous farmers, I was close to the action. I spent several afternoons tramping farmers' quarter sections, followed up with an invitation to supper. And once in a while someone paid me. It was a way of getting to know a farmer while seated at his kitchen table.

Then the business of converting cellulosic materials to gasoline from ethanol took the shine off agroforestry, especially after funding for this program shrank. For me, I found converting wheat to ethanol dumb.

Ω

By September 2003, Marilyn and I pulled the pin, and settled down on the Sunshine Coast, a ferry ride west of West Vancouver. I continued to do my forestry consulting in Prince Albert until November 2010. I left my soul in Saskatchewan, so there was always an excuse to find myself back in God's Country.

November 2003: Sturgeon Landing timber salvage

Dale Reid, Chief Ronnie's former executive assistant, now a PBCN development assistant-at-large (carpetbagger) and I had inspected the recent plow wind damage at the Peter Ballantyne Cree Nation Sturgeon-Weir reserve, IR 184F on August 16. That beautiful stand of white spruce trees, over 800 ha in area, which grew on a rich fluvial bench, had needed to be logged for years. It was overmature and was 138 years old – way beyond its 'best before' date. The band had mapped and timber cruised this stand of timber previously. Another remarkable performance by forestry assistants from Pelican Narrows. In fact, Henry Morin had done some previous logging on the reserve close to the shoreline of Namew Lake. The decked logs were left to rot.

The Sturgeon-Weir band members, as well as the residential school, were originally located in housing on a cleared grassy bench running up the west edge of the Sturgeon-Weir River, not far from its mouth. However, there was no vehicle access. This ended at the River on the Manitoba side. From there, a wire rope suspension bridge suitable for foot traffic and light freight provided access to the houses. The school burned down in the '50's, leaving lost graves now under tree and shrub canopies.

Gertie Budd was the PBCN counselor represented the sixty-six band members on the reserve in 2003. By this time Indian Affairs had constructed new housing units and moved the band housing over to the accessible Manitoba side. It became an informal attachment to the indigenous community of Sturgeon Landing. Gertie had a 6.5 hours trip each time she had to leave on band business. To get to the band's head office in Prince Albert, she needed to go east to Manitoba Highway 10, then south through The Pas. She then drove into Saskatchewan on Highway 9 turning onto the Kelsey Trail. She stopped in Nipawin for a bite to eat and gas for her Ford 'dually' pickup. Her travel allowance was modest. The last time I saw her truck in the parking lot the tires were chewed up. Both spares were flat.

After fighting through windfalls and pockets of standing timber, Dale and I visited the Tolko Industries sawmill/pulpmill complex at The Pas. Yes, they were most interested in a timber salvage opportunity given the timber was of such high quality. The oversize logs would be shipped to Saskply at Hudson Bay in return for an equal volume of Saskatchewan sawtimber. I left them the timber cruise maps to look over.

A portion of the blowdown on the Sturgeon-Weir reserve was an excellent grade of poplar. Tolko was also interested in salvaging this as well. We left Tolko to do the 'due diligence' and come up with a rental fee per cubic metre for the timber volume removed. Dale's job was to brief Gertie on this opportunity, including jobs for band members. In turn, Gertie's task was to brief Chief and council and obtain their consent for such a project. I went home and waited for the wheels to turn.

After returning to Prince Albert in October, Gertie and Dale asked me to generate a forest operations plan for submission to Indian Affairs, Prince Albert office. Also, Tolko requested that I get all the conditions and a permit from the federal fisheries officer in Prince Albert for accessing Namew Lake to build an ice road. The Company would use this for hauling the timber from the reserve over to the Manitoba side, thence to the mill. Much red tape, but this was necessary if we were to maintain environmental stability within the area in question. Of special interest was the preservation of a heronry and a grotto used by band members to renew their spiritual spectrum. The former was located in a grove of big poplars at the lake edge, while the latter was hidden beneath a large rock shelf.

I did locate the heronry on the proposed forest operations map as an exclusion. However, I left the grotto off the maps since it was a sacred place. Gertie started with the timber salvage planning phase. She had received a BCR (band council resolution) approving the project. Dale negotiated a rental fee of $5 per cubic metre for the timber scaled in the sawmill yard. Tolko would remit the monies biweekly into an account set up by PBCN administration. One of my jobs was to reconcile the biweekly Tolko scale summaries with monies received.

I received the permit and conditions for the ice crossing from the federal fisheries office. I ran into some complications with the forest operations plan approvals. Indian Affairs had turned this over to the Canadian Forest Service office in Prince Albert. This agency wanted to strengthen the section dealing with the post-harvest forest renewal plan. I had to add a schedule that included the source of the white spruce seed to be used in the Prince Albert forest nursery. This meant approval from the province (Saskatchewan) to supply seed from its tree seed inventory suited the area logged. Then a delivery date from the nursery to coincide with spring planting, then a plan to pick up the seedlings from the nursery and deliver them to the planting site.

I also had to calculate the number of seedlings needed and the planting sequence. Gertie chose to find tree planters to plant 250,000 spruce seedlings in May 2005 and again in 2006. I set up the arrangements with nursery to grow the trees, then with the band to pay the invoices through the Indian Affairs trust account. This would originate with a $5 per cubic metre payable by Tolko as a condition within the Indian timber regulations. No Indian timber is to leave the reserve without timber scaling and the payment of Crown timber dues. Tolko negotiated with Indian Affairs to scale the loads of logs at its mill then remit the monies to Indian Affairs in trust.

Figure 27 Gertie Budd supervising the movement of logging machines across the ice, January 2004.

As soon as the ice thickened to its regulation loading allowance, the logging equipment was brought across the ice and a camp set up on the reserve. Tolko moved in three of everything needed to get that wood out by breakup. The Company brought in a trainer and, before you knew it, nine band members became adept with the controls on several machines. They made good money.

Not only that, but Tolko used some of its own seedling supplies to train band members to plant trees on the logged over ground after breakup. Our own tree seedlings were just starting to germinate back in the forest nursery. Before returning home for the summer I made sure the fisheries inspection after break up was completed successfully. Straw bales and drainage cloth had been laid down at both ends of the ice road crossing to eliminate silt from flowing down a melting road on to the lakeshore. It did not help that SaskPower let more water through its Nipawin dam during the peak logging season and flooded the ice road. This froze brakes to the point that a cat had to tow machines to crack the ice in the brake drums. On my way up to Creighton one afternoon, I drove with brake pedal depressed in order to melt ice in the brake drums. What a performance!

Before I left home in March I hired a forestry technician from Prince Albert to help with the spring tree planting. Tolko organised the training program. We ordered sufficient planting tools, bags, safety gear and accessories for a fifteen person planting crew. He also ribboned in new access trails for the following winter's logging. I left the band administration with a record of what Tolko had paid, both for Gertie Budd's rental fees and the Indian Affairs trust fund.

Ω

The startup for the next winter's timber salvage was smooth. I had the necessary fisheries permit in hand. The 2004-05 winter logging session ran well. Except for periodic logging inspections and the audit of timber volumes produced and paid for, I now concentrated on preparing the spring tree plant project.

Then everything fell apart. I could not get the seedlings paid for by the band. So 150,000 white spruce nursery stock had to find a home elsewhere. The PRT nursery manager finally found a customer in the region.

Tolko logged and bought 193,000 cubic metres of spruce over those two winter seasons amounting to $1.9 million owing to Sturgeon-Weir reserve and PBCN administration. Indian Affairs scooped the lot to pay down Peter Ballantyne debt. To hell with the reforestation plan – even though they set this out as a condition of approval for the timber salvage. Unscrupulous government at work. The Honourable Andy Scott, Minister of Indian Affairs and Northern Development in those days does not appear to be as honourable as we had assumed him to be. He supported rape and pillage of a precious timber resource.

Chapter 30:

Observations on running a small business while at the mercy of the banker.

Both IFFS and Silviba needed an even cashflow over the years:

International Forest Fire Systems Inc. IFFS had six cost centres in its diversified operations as a means of survival. TD Bank was the main banker most of the time. Each spring it received our consolidated business plan for consideration in arranging a loan for that year's operations. The loan-seeking process always started on the ground floor in a glassed-in cubicle. Much glad handing, coffee and idle chit chat. At the last moment, the bank officer would accept the year's pro forma operating forecast and promise a quick response.

As the days and weeks went by, matters grew more critical each day as the company prepared to start on its spring contracts. The further up in elevation our loan request travelled, the less attachment the official on the ground floor had to us and our wants.

As time progressed we divided the company into divisions and went after funding at different banks. The bankers for the different IFFS divisions maintained connections with one another.

The Bank of Montreal went for your guts if there was a default. After Bob Henderson got my wife and Mrs. Poliquin to cosign a huge loan to purchase the Aero Commander, we found out the problem lay with the Canadian Imperial Bank of Commerce. But BMO got our personal assets first. One of our directors was the manager of a CIBC bank in Victoria. Oh, the banking world.

Silviba Services (Ltd.): Silviba, the contraction of 'silviculture' (the growing of trees) and '(f)iba', the mowing of trees (logging), started life as a division of IFFS in 1977. It was rejuvenated as a Saskatchewan registered company in May 1984. Silviba's first bank in Prince Albert was the newly arrived Bank of BC, with premises in the Gateway mall. But it folded soon after, and I found a new banker, TD Bank, shortly thereafter.

Eventually this bank branch moved into new premises at thirteenth Street and Central Avenue in Prince Albert, right across from my office. The bank management's business loan authority was transferred to Winnipeg. Silviba became an orphan in its own house. It had few physical assets – mostly those of an intangible nature, thus useless as collateral.

I remember that stressful time between banks. I cashed in RRSPs and kept searching. In time, I found the Prince Albert Credit Union (PACU). It was 'my cup of tea'. It was no less diligent, but communication was easier. I used Denis as a 'director at large'. The one feature I loved about PACU was that loans were adjudicated in house, not in Winnipeg or Toronto. For small businesses, this is a huge advantage.

Eventually, the increased business accounts at PACU created the need for more staff. These came from banks, and they brought some of their banks' bad habits with them. Both the Royal Bank and CIBC had purged their small business portfolios, and thrown their small business account holders to the wolves. This was simple to do – just call in the loan. They caught some larger companies in the same net as well. Never trust a bank.

Silviba used the PACU to the end. At some stage, PACU had built brand-new premises at the corner of 28th Street and 2nd Ave West. Monica became our account manager. She could say 'no' easier, but she was the necessary brake at times. This energized the process of keeping on top of our invoicing.

The Business Development Bank in Saskatoon started out as a good listener in the '80s. Latterly, it took on the air of a precocious protégé with all the polished glass and stainless steel decoration. Later, my eye caught a reference to its association with Bombardier in a Globe and Mail newspaper. The Bank seems to have lost track of its mission.

Silviba grossed $2.2 million dollars between 1985 and 2003. It expensed $2.1 million, but some of this was depreciation.

PART THREE

Early Forest History in the Prince Albert District, Saskatchewan District, NWT.

Based on the 'Balance of Probabilities.'

Ω

This completes a long-standing personal commitment to the Prince Albert Museum and Saskatchewan Archives Board.

Ω

I am asked why I included this section as part of 'A forester's log'. My answer is that this was the outcome of a significant portion of Silviba's workload from April 1997 until late 2009. The result of the analysis of archived records, matched up with forest ecosystem evidence on the ground shows that 3.5 million logs or more originated from Indian lands. The logs had no timber mark to denote ownership. They were shipped to Prince Albert Lumber in Prince Albert as part of the spring river drive and not recorded. No timber dues were paid to Indian Affairs in trust for the two reserves impacted. The annual river drive connected all log supplies in the bush with the sawmills in Prince Albert.

Ω

Heather Patterson prepared a GIS version of the timber limits north of Prince Albert existing since 1886. The inputs came from the individual timber berth files. I hope to leave one map at the Prince Albert museum, a second at Saskatchewan Archives Board and a third – digital version – on the proposed Silviba website. Reference: hpatterson@forsite.ca

Chapter 31:

Before Prince Albert Lumber Co. Ltd.

Wherever you go in southern Canada, flat land attracts farmers or emerging communities. Add in an ingredient that brings people together in one place, you create an overlay of human endeavour. In Prince Albert's case, it started with the fur trade. The first White to travel through the area that is now Prince Albert was Henry Kelsey in 1692. The first establishment in the area was a trading post set up in 1776 by Peter Pond. [88] James Isbister, an Anglo Métis employee of the Hudsons Bay Company, settled on the site of the current city in 1862. He farmed there until 1866, and was joined by families who called the site Isbister's Settlement. The community was founded in 1866 by Reverend James Nisbet,[89] a Canadian Presbyterian church minister who had come to set up a mission for the Cree. Nisbet named the community after Prince Albert, Queen Victoria's Consort. The party arrived at what is now the Prince Albert area on July 26, 1866 – the site originally an Indian campground called *kista-pinnanick*, which translates into 'meeting place.' (Prince Albert Daily Herald, March 14, 2014).

Woodland Cree people, Plains Cree, and Assiniboine populated the region long before the birth of Prince Albert. Publication No. 3200 from the Prince Albert Model Forest is a 'must read'. Otherwise, how can we begin to understand the aspirations of our neighbours?[90]

Indian bands had used the North Saskatchewan River as a highway for centuries, had buried their dead where the Prince Albert Forestry Centre, complete with ghost, now stands. This area was a trading centre for Indians passing up and down the River, as well as for the surrounding Indian communities. When you add in the travel corridors between the Forestry Centre and the Churchill River via La Ronge and Montreal Lake, the trading area was significant. Even the Athabasca Indians became involved as they pushed the Woodland Cree south and beat them up at Hanging Heart Lake (oral history). By the early 1800s food was becoming a real issue for all bands within which is now the Prince Albert Grand Council. The fur trade and climate change impacted on the health and welfare of all First Nations in the region. This included reduced sustenance and white man diseases. The Cree James Bird and his family were the first true immigrants in the region and hung their hats at Montreal Lake.[91] Our first settlers.

The entire District of Saskatchewan, Northwest Territories was at the tail end of a mini ice age from 1750 to 1920[92]. Winter ice four feet thick on the lakes, humid wet summers and heavy snowpack provided ideal growth conditions for the carpet of white spruce forests north of Prince Albert. These grew all the way up to the top end of Montreal Lake. No insects to chew all the leaves off, no bark beetles to bury their way inside the trees and kill them off. No fires to wipe out entire forests. After adding in the good health of prairie grasses, you have the makings of a push for white settlement.

However, settlers needed transport. Thus, railways came into the mix. One enthusiastic group even had a right-of-way surveyed in from Saskatoon all the way up past Montreal Lake. Railway construction includes the need for large supplies of rail ties. The first forestry efforts in Prince Albert district included making rail ties from conveniently close-by jack pine stands in what we now call the Nisbet and Fort a la Corne forests. Shortly, these also became the source of fuelwood as well for those new homesteaders in Prince Albert. The federal timber agent gave out timber permits for 200 cords at a time. He also issued permits for small sawmills to supply settlers with building lumber.

The emergence of forest regulation and timber supply
A) On lands under the jurisdiction of the Department of the Interior

The British North America Act in 1867 resulted in Canadian government administration. The first Dominion of Canada Forestry Act became law in 1870. Coincidentally, the British government handed over Rupert's Land and the NWT possessions to Canada, to create an expanded Northwest Territories jurisdiction. In 1875, the government of Canada brought in the amended Northwest Territories Act to remove duplication and discrepancies in administration. In 1882, the District of Saskatchewan became one of the four new administrative districts. Administering the Dominion Forest Act and timber regulations became the responsibility of the Department of the Interior. For Saskatchewan district, the regional office was Winnipeg. Prince Albert became an administrative subdivision that had responsibility for local settlement, agriculture and forestry. In January 1903, the timber officer in Prince Albert complained strenuously to his superior in Winnipeg that he was too busy processing homestead applications for 423 new settlers to worry about timber administration.[93]

The federal government regulated the forest industry in Saskatchewan until 1930, when the Natural Resources Transfer Act ceded all Crown lands, including forest lands, except Indian reserves and National Parks, to provincial jurisdiction. The first Dominion Forestry Act in 1870 created timber regulations that defined the rules for cutting federal timber. After 1880, timber leases up to 50 square miles could be applied for. They were subject to an annual licence fee at $2 per square mile. After the timber was sawn up, the government charged an *ad valorem* royalty per 1000 board feet (MBM) at 5% of sales.

These licences were renewable annually. The applicant surveyed the timber berth boundaries using a Dominion Licensed Surveyor (DLS) before receipt of the annual licence. The owner renewed this every four years, provided all dues and fees were paid beforehand.

But in 1889, the government introduced timber dues at so much a 1,000 board feet (MBM) to increase Crown revenues. This increased to $1/MBM[94] or more over time. Lumber sold at $17 to $18/MBM. The government also taxed shingles, lath and other products. This policy resulted in lumbermen often underestimating

389

and under reporting the timber volumes sawn to reduce these royalties and fees. The timber berth system led to creating companies with vast timber sources, such as Prince Albert Lumber Co. Ltd, to come.[95]

After logging started, an operator had to present a map at the end of each winter that sketched the area logged. Before he could start logging again, he had to send in the annual licence fee and land rent. He also presented sworn quarterly returns to record the volume of logs removed from the licence. There were 840 square miles of prime spruce timber between Prince Albert and the south boundary of the Sturgeon forest reserve, now Prince Albert National Park. Licensees surveyed and registered 28 timber berths. The gross timber volume available for harvesting amounted to 3.22 million board feet of white spruce sawlogs per square mile.[96] Estimates gained from fifteen years of sales data originating in the Dominion timber berth files showed 2.64 billion board feet of available spruce lumber according to the Prince Albert Lumber Co. between 1905 and 1921.[97] This amounted to 53 years of timber supply – just in the area under licence. And did not include the Indian reserves.

Another section in the legislation allowed for annual timber permits. These permits triggered the wood supply for many small sawmills that produced building materials for the settler market. Often the settler would hand over his cutting rights to the mill owner.

In 1907, Forestry Branch, Department of the Interior, took over responsibility for forest reserves. In 1912, Timber and Grazing Branch assumed administration of licensed timber berths. That legislation also delineated the Sturgeon Forest Reserve, the forerunner of Prince Albert National Park.

B) On Indian lands

At the same time, the emergence of Indian treaties also impacted the Prince Albert timber supply, or wood basket. The Indian Act of 1876 led to Treaty 6 with Plains and Woods Cree, signed at Fort Carlton in August 1876. Following this, both Montreal Lake and Lac La Ronge Indians signed an adhesion at Molanosa on February 11, 1889. Revised Indian timber regulations for the NWT became law in 1887. Two new reserves with commercial timber potential become the essence of this saga: The William Twatt band (future Sturgeon Lake First Nation) signed on at Fort Carlton on August 23, 1876. The Molanosa adhesion produced the Little Red reserve created in 1897, just to the west of Northside, Saskatchewan. This was a shared reserve eventually populated by band members from both Montreal Lake Cree Nation and Lac La Ronge Indian Band. The Department of Indian Affairs faced a depressing fact – Indians were starving and needed encouragement to grow their own food. The Prince Albert Chamber of Commerce deeply resented the location because of its grazing potential, now under the jurisdiction of Indian Affairs. In fact, a Department of Lands surveyor had previously run cattle there in the summers. He was miffed and joined in the howl directed at T.O Davis, the Liberal MP, later senator, for Prince Albert. He, in turn, howled at Clifford Sifton, the federal minister responsible for the Department of the Interior. The latter was a conniver of significant reputation.

The Department of Interior correspondence is housed within catalogue NR 5 at Saskatchewan Archives Board, University of Saskatchewan. This reeks of 'insider trading' connected to land accumulation and conveyancing in the Prince Albert district. The inference from examining this correspondence is that George Burn, the bank of Ottawa manager in Prince Albert was in the thick of it[98]. He moved to a bank position in Ottawa sometime soon after 1900. It was his signature on cheques drawn on the Bank of Ottawa at head office that paid for timber licence fees[99]. Then, Adamson, the federal MP for Rosthern and J.G. Turriff, MP for Assiniboia East (a merchant) became the second and third 'musketeers'.[100] Together, they formed a company – Canada Territories Corporation.

Someone paid the new agriculture instructor at Little Red reserve $100 (1897 dollars) to buy the surrender to this new reserve while the ink had barely dried on the Treaty adhesion. In that first few years, a handful of Montreal Lake Indians had begun to scratch a living growing potatoes. They signed an agreement to surrender the reserve. However, they had no authority to do this. The local Anglican minister became involved at the request of chiefs and councils. The matter dissolved shortly after[101]. However, the relationship between the two bands as they shared the same 56.5

square miles was not one of mutual comfort. Subdivisions created later by Indian Affairs ameliorated some of the stress.

Chakastapaysin reserve, located between St Louis and Red Deer Hill, evaporated as well. Dispossessed Métis and Indian fighters moved in after the Battle of Batoche. It had occupied 34.5 square miles of potential homestead land[102]. Although, some band members had remained peaceful, they felt severely threatened if they did not join the Rebellion. Anyway, T.O Davis and his musketeers engineered the surrender of this reserve, then subdivided it and sold the land to homesteaders.[103]. According to James Smith band research, it was Turriff that fronted the money to buy Chakastapaysin lands.[104]

Indian timber harvesting regulations after 1878

The administrative imperative for Indian lands in the west revolved around agriculture. Forestry was barely mentioned. The Indian Affairs sessional reports received annually by Parliament were bereft of any reporting of timber logged. Lots of hay put up, many cattle recorded, some crops produced – but nothing, nothing on trees, logs or Crown dues collected in trust. This became glaringly apparent. First, Sturgeon Lake First Nation brought a timber surrender issue to the Indian Special Claims Commission in 1997. It dated back to 1906. Second, the combined Montreal Lake Cree Nation and Lac La Ronge Indian Band (or Little Red), brought illegal timber surrender issues dating back to 1904 to the same Commission in 2003.

In summary:

- No one can remove logs from an Indian reserve without a licence from the minister.
- No licence will be issued without approval from chief and councilors.
- An approved licence contains permission for the removal of a stated volume of timber, or used within the boundaries by band members, logged and scaled.
- Each application for the removal of logs will be accompanied by an advertisement in the newspapers

pointing to the operator's intention and the area in acres to be logged. Plus a bonus bid for the timber removed.

- In his application, the operator will include a cheque for the first year's licence of $5 per square mile. He will also include a cheque to cover the bonus bid in $/1000 board feet of lumber produced.
- He will agree to provide Indian Affairs with sworn statements of the timber volume scaled and sawn every quarter.
- He will pay the Crown dues designated.
- He will apply for a renewal and provide a cheque for the next year's licence renewal.
- The application will also include the log specification and stump height designated.
- A statement of the timber volume to be logged by species
- Penalties for non-compliance include paying three times the dues owing, or at some other rate as determined by the minister.

Agent Keith at Fort Carlton was responsible for land administration on reserve lands. His office was at Fort Carlton and his trips to Sturgeon Lake and Little Red covered 45 miles one-way, on horseback.

He was fortunate in that a wagon road, the Prince Albert to Montreal Lake trail, complete with a bridge across the Sturgeon River at White Star passed right through both Sturgeon Lake and Little Red reserves – he had good access. But so did the Prince Albert loggers.

A digest of a detailed investigation to discover the quantity and location of spruce sawlogs stolen from both the Sturgeon Lake First Nation, and Little Red reserve from 1904 onward follows next. In April 1997, Sturgeon Lake First Nation asked Silviba to investigate the possible theft of spruce sawtimber in 1906, and the volume removed in trespass. This work commenced in 1997 and ended with a settlement in in 2001. Then in 2003, the two First Nations residing on Indian Reserve 106A, Little Red, asked Silviba to investigate timber thefts between 1904 and 1911. This latter action remains in the courts. This has tied my hands somewhat.

The Prince Albert wood basket, 1900 plus

Three gentlemen in Prince Albert Dominion grabbed up timber licences, referred to as Dominion timber berths (DTB). The Prince Albert timber agent (source: file NR5 A1 a9, Saskatchewan Archives Board) administered the Timber Regulations.

James Sanderson	10 berths totaling	87.14 square miles
William Cowan	7 berths totaling	59.25 square miles
George Burn	11 berths totaling	249.75 square miles
Total		396.14 square miles

They intended to log these berths clean of all merchantable timber, and then walk away. They would revert to the Crown, which would then sell the denuded lands to settlers. Probably through the Canada Territories Corporation (the three musketeers).

The amassed timber supply from these berths is estimated at between 13 billion and 26 billion board feet. This was exceptional volume of high grade white spruce sawlog material. It was all accessible to Prince Albert lumbermen, who together logged up to 15 million board feet a year, if they could sell it.

Sanderson's sawmill was on TB 9 block 3 just north of the middle of Sturgeon Lake. Cowan started out on the south bank of Sturgeon Lake according to Hanna Kingfisher.

Telford's sawmill was on the east flat in Prince Albert. The brothers logged from 'limits on Sturgeon Lake'.

Shannon's mill on Bell Lake existed to the west of the northwest corner of little Red reserve. His timber came from limits on Little Red River north of the IR 106A north boundary.

George Burn was the local bank of Ottawa banker, moving soon to Ottawa. He was closely linked to Senator T.O Davis and local MPs in business matters. He controlled timber rights on 250 square miles immediately north of Prince Albert.

Schedule B

Statement showing Saw-mills in the Prince Albert operating under government Licence during the year ending June 30, 1904

Name of Owners	horsepower and type	Capacity per 12 hrs	Commenced operations	Description of timber	Where cut, location of Limits	Quantity of lumber manufactured in year	Quantity of lumber sold	No. of shingles manufactured	no. shingles sold	Date of last return & total of returns made
						MBM *	MBM*	M pcs		
James H Sanderson	225 Steam	36,000	1888	Spruce	North of Sturgeon Lake	3,051	2863	478	425	Jne 30, 1904
W.Cowan Co	35 Steam	30,000	1890	Spruce	Up little Red R	4,389	3,733	648	648	Jne 30, 1904
Shannon & Co	35 Steam	14,000	1899	Spruce	Up Little Red R.	2,188	1,156	116	116	Jne 30, 1904
Telford and Co.	264 Steam	75,000	1902	Spruce	On Sturgeon Lake	5,680	7,752	319	275	Jne 30, 1904
					Totals	10,919	15,504	1,561	1,464	180

Crown Timber Office

Prince Albert, July 11, 1904

J.W. HANNON

Crown Timber Agent

Copied from   Source: 4-5 Edward VII., A, 1905

* MBM = 1,000 board feet

Figure 27: Copy of table contained in the Prince Albert timber agent's report, July 11, 1904.

COPY

"DEPARTMENT OF THE INTERIOR

4 – 5 EDWARD VII., A. 1905

No.13

REPORT OF THE AGENT AT PRINCE ALBERT.

Dominion Lands Office

Prince Albert, June 30, 1904

The Commissioner of Dominion Lands.

Ottawa, Ont.

Sir. – I have the honour to submit the report of this agency for the fiscal year ending today.

Great and unusual difficulties affecting the branch line of railroad serving this district have made this season much less favorable to immigration than was fully expected. Largely from this cause the homestead entries have not reached the high water mark of last year, and there is a consequent falling off in the total revenue. I would note, however, that in spite all the difficulties the homestead entry are more than a half greater than those of two years ago, being 1,636 as against 1009 for that year, and the same steady and gratifying increase is seen in total revenue, which is this year $36,304.68, as against $28,958.81 two years ago.

The British immigration has been marked during the past year, and several large parties of a very superior class have been distributed throughout the district. There is also a considerable French settlement being made east from Duck Lake, while the Norwegian settlement at Glen Mary and German catholic settlement on the Hoodoo plains are being steadily increased incomers from the United states.

The present crop outlook is the best seen here for many years, and if realized it will not only put new settlers on their feet but should stimulate increased settlement next year under, I hope, more favorable transportation conditions than were experienced this year.

I am glad to note that the Canadian northern Railway has now entered the district. Its progress continues to be watched with the greatest interest, and it is of the utmost moment to all. The increase of railway facilities is the great and pressing need for the development of progress in this district.

In the following statement of the work of this year, the increase of the correspondence will be noted, as well as the marked increase of cancellations, which are more than double those of last year.

Letters received		8.479
Letters sent		6.937
Homestead entries granted		1.636
Homestead entries cancelled		636
Hay permits issued	47	
Timber permits issued	1.262	
Application for patents received		273

Your obedient servant.

J.W HANNON

Agent of Dominion Lands"[105]

Ω

Annual reports presented as part of the Sessional Papers for the period displayed further lumber statistics. These follow next. It is important to realise that all logged timber and lumber produced under the Department of Interior timber regulations was based on sworn quarterly returns to the federal timber agent in Prince Albert.

A summary of spruce volumes logged near to Prince Albert according to the timber agent's annual reports until 1910 follows:

Year	Sanderson	Cowan	Shannon	Telford	Sturg Lk Lbr	Other	P.A. Lbr	Total
LUMBER PRODUCTION, PRINCE ALBERT DISTRICT, DEPARTMENT OF INTERIOR.								
Lumber manufactured from stationary mills, MMBM.								
1903/04	3.1	4.3	2.3	5.6		0	0	15.3
1904/05	2.6	4.6		5.4	3	0	0	15.6
				** mill burned down May 1905				
1905/06	4	2.9			4.6	3.5	4.2	19.2
1906/07	2.7	3.2			4	3.7	14.9	28.5
9 months								
1907/08		4.5			1.25	6.8	37.6	50.15
15 months								
1908/09								
1909/10						8.6	32.8	41.4

Note: Information supplied by Alexander Dietz from sessional papers, U. of Saskatchewan.
"Other" sawmill volumes are not tributary to P.A. Lumber wood supply area.

Silviba   sturg\anrpt.xls

Figure 29: Summary timber volume converted to white spruce lumber, 1903-1910 in Prince Albert. Source, catalogue NR 5, Saskatchewan Archives Board

The 'other' category is independent settler timber permits assigned to Prince Albert Lumber Co. Ltd. Volumes are in millions of board feet.

Sturgeon Lake Lumber Co. Ltd, 1904-1907

Sturgeon Lake Lumber Co. Limited sawmill ran eight miles north of Sturgeon Lake. A.J. Bell was president. William Shannon provided the sawmill operations expertise for this enterprise. The mill was partially financed by Canada Territories Corporation. Its principals were three co-investors, Messrs. T.O. Davis MP, A.J Adamson MP and J.G. Turriff, sometime Land Commissioner. Funds came through The Western Trust Company of Winnipeg.

The Sturgeon Lake Lumber Co. succeeded the Bell and Shannon mill, found within Dominion timber berth 848 (TB 848) and just north of the northeast corner of TB 598 (See Map, page 14). Cyrus Carroll, DLS, describes both the mill and timber within TB 848. Annual reports of the Department of the Interior, Prince Albert district show that Sturgeon Lake Lumber Co. produced 3 to 4 million board feet of white spruce lumber yearly between 1904 and 1906. This dropped to 1.25 million board feet in 1907/08. Of course, this originated from logs licensed under the Department of Interior timber regulations, and governed by the timber agent in Prince Albert. A

Department of Interior audit in 1911 captured logs originating from Little Red reserve.

This mill was first-class: headsaw, resaws, steam chest for warming up frozen timber and living quarters for the crew. It produced lumber of several dimensions as well as lath[106]. Remember, in those days, lath was used on all walls and ceilings as a base for the plaster. They loaded rough lumber produced on to sleds and freighted the train in to Sanderson's planer mill in Prince Albert before the Montreal Lake trail became impassable in breakup.[107]

Figure 30: The rough lumber haul from Bell Lake to Prince Albert. Saskatchewan Archives Board and Bill Smiley collection.

To this point the Prince Albert forest industry was a collection of small sawmills totaling about 15 million board feet of planed and dried spruce lumber each year. Timber berths within the Sturgeon Lake wood basket supplied the logs.

In the following chapter, Prince Albert Lumber Co. Ltd enters the picture.

Ω

Chapter 32:

Prince Albert Lumber Co Ltd. (PAL), 1905-21

TR's Note

All photographs in this chapter are stored at the Regina branch of the Saskatchewan Archives Board. They are part of photographer James' collection. He worked out of Prince Albert. I believe he took many of these in February 1907. We know this was a very cold winter since Mr. Elliott had told the Prince Albert newspaper cold weather delayed logging. He also told the newspaper they had brought in seventy million board feet of logs. However, the ice on Sturgeon Lake would have set early and provided the 4 feet of thickness needed to support the timber decked there. According to measurements taken from James' panoramas of logs decked on Sturgeon Lake, the ice on Sturgeon Lake supported at least 50 million board feet of log storage capacity. The reader must realise the facts displayed in Part Three are the result of intimate familiarity in the operations of the forest industry. Loggers needed to deliver logs at least cost, and the sawmills needed to maximize the returns from lumber sold. The photo displayed in Figure 48 can only have been taken in late 1910 after construction of the spillway on Little Red River. The newly-arrived river drive engineer installed new dams and weirs on the Little Red River. Photographer James drove his two-horse sleigh up the Montreal Lake trail to a point south of the Bell Lake confluence.

The dilemma is: how do we discover total lumber production from the Prince Albert wood basket when <u>only</u> the Crown timber berth logs were measured, recorded and accounted for?

Ω

In some years, PAL only reported 25 million board feet of lumber production. This, for a mill with an annual capacity of 50 million board feet? With a reported log production of 70 million board feet in 1906-07? The balance of its harvest came from Indian land immediately accessible to Prince Albert through both the Sturgeon and Little Red Rivers. It did not come from the moon.

Ω

We finally get to lay out the fuzzy facts behind the biggest forestry secret in Saskatchewan: Prince Albert Lumber Co. Ltd and Ladder Lake Lumber Company (Big River). PAL was the second largest sawmill in the British dominions in 1905. Ladder Lake Co.at Big River was the largest. However, the latter is not studied here. It kept burning down; it was designed to produce twice as much lumber as PAL (100 million board feet a year). None of the wood flow within the Prince Albert wood basket was accessible to Ladder Lake, given the character of the transport pattern. All the waterways pointed to Prince Albert. However, one timber berth straddled the height of land between Sturgeon Lake and Cowan Lake. Both outfits removed logs from it, but this was all recorded and accounted for. Albert Mattes, on the Prince Albert Lumber Company payroll, managed both mills.

Martha McCartney describes how the Winton Brothers moved into Prince Albert from Wisconsin in 1905.[108] Americans had emptied the wood basket in Minnesota and Wisconsin, and with a zero timber inventory, they shut down and moved into western Canada.

The Telford Brothers sawmill in Prince Albert burned in May 1905, one brother was too sick to keep working and the second brother wanted out of the business – according to Prince Albert museum sources. So, the Wintons rebuilt the Telford mill (A mill) and built a brand-new B mill. The new company became Prince Albert Lumber Co. Ltd. (PAL). It was gazetted and registered in Ottawa.

This was a major development for the times – "the largest and most complete mill between Rat Portage and the British Columbia"[109]. The Winnipeg Free Press in its March 18, 1906 edition devoted a 2-page spread advertising the economic attractions of the Prince Albert district. It included the new PAL mill – a capacity of 50 million feet of spruce lumber a year as well as 12 million laths.[110] Rat Portage was what we know now as Kenora, Ont. This was matched by a spruce sawlog supply from "850 square miles of good spruce under license tributary to this point (*sic. Prince Albert*)."[111] Since a new sawmill of any scale needs logs urgently, this became the task of PAL's logging superintendent, Mr. Elliott. Department of Interior records show that PAL manufactured only 4,267, 000 board feet of spruce lumber in 1905/06. In those early Sessional papers a year started on July 1 and ended the next June 30. The source had to be Telford's log yard inventory given the mill came on stream just six months after Telford's last river drive (May 1905).

Figure 31: The new 'B' mill, Prince Albert Lumber Co Ltd, 1906. Note the logo (Winton). Saskatchewan Archives Board and Bill Smiley collection.

In the June 12, 1906 edition, The Prince Albert Times Weekly newspaper reported that a new general manager, Albert Mattes, had arrived in Prince Albert. The mill ran at full capacity by September 12, 1906. A fire in B mill in December 1906 held up production. It was running again shortly after. On February 28, 1907, this newspaper made reference to a volume of 75 million board feet logged for the winter[112], of which 60 million board feet of lumber would become lumber that year (1906/07). Yet from Table 1A that follows, PAL only reported a volume of 24.7 million board feet removed from timber berths. Sturgeon Lake Lumber Company supplied much of the timber inventory that winter. The British Columbia Construction Co did the logging under contract. The only accessible unrecorded timber volumes came from nearby Indian reserves. The timber berth records show only the reported timber volume logged – a significant shortfall. The Indian reserves, Sturgeon Lake First Nation, IR 101A, and Little Red, 106A held the closest blocks of timber to Prince Albert. Of course, this was stolen timber, free of dues or annual rental. Loggers love free wood. Technically, we refer to this as a trespass cut. I conducted investigations on several occasions during my career. It is always the same: bigger, more valuable timber on the other side of the line.

Timber supply

The Sturgeon and Little Red Rivers became the focus of the spring log drives through Sturgeon Lake and Little Red reserves. Early loggers had dammed Sturgeon Lake at the east end[113]. After freeze up it served as the landing place for all logs originating from both the shores of Sturgeon Lake. Skid roads connected the Lake to timber operations conducted in nearby forestlands. Log trains transferred logs from the bush on to the ice. It was easy logging, both on the Indian reserves, and on close by Dominion timber berths. This is observable from James' photographs.

Figure32: (R –B11417). A tractor train arriving on Sturgeon Lake. Saskatchewan Archives Board

Lumbering Scenes near Prince Albert, Sask.

Figure33: (R – B1039 – 3). Logs decked on the Little Red River. Saskatchewan Archives Board (Bill Smiley collection?)

One believes these log decks were placed on the Little Red River south of Bell Lake. This runs right through the reserve and would become the east access to Indian timber. PAL had built a dam equipped with sluice gates upstream. Photo in the Regina Archives files.

Ω

With ice four feet thick, PAL decked 50 million board feet [114]on the ice, ready to spill down those rivers in April. Peter Pannabacher, the PAL accountant, did a study of log losses on the way to Prince Albert with the spring river drive in May 1910. He found that average log losses were 7.5%. For TB 598 on the western boundary of Little Red reserve, log losses amounted to 25% of the gross volume logged. (Information from the PAL files archived at the Saskatchewan Archives Board in Saskatoon).

Large tracts of high quality Indian timber lay in front of the Dominion timber berths signed over to PAL by George Burn and others. Loggers grabbed these first because they were so accessible.

Table 1 A. Reported Lumber volumes

STURGEON LAKE WOOD BASKET - ACTIVE TIMBER BERTHS 1905 TO 1919

Text Table 5 — Reported lumber production, Prince Albert Lumber Co.

Faith Sort Final.xls

TB	1905/06	1906/07	1907/08	1908/09	1909/10	1910/11	1911/12	1912/13	1913/14	1914/15	1915/16	1916/17	1917/18	1918/19	Total MBM
245	.	7,000	10,676	664	.	.		.	.	.	.	.	9,594	23,661	51,595
320														28	28
474				4,811	6,642		18,894	10,647	11,991	2,986	6,658	16,966	11,239	7,015	97,868
598					19,430	5,932									25,362
633						6,705	6,701	3,753		6,534	4,684	210			28,587
669									20,336						20,336
691			1,362					2,896				2,595	8,215		15,168
698			168	1,134				5,660				2,317			9,299
708/402	2,228	2,479	693									3,448			8,748
710	909	1,118	3,909										7,703	6,558	20,197
714	1,130														1,130
729						3,232	8,105								11,337
766									2,057	12,202	154				14,413
801										4,317					4,317
868								16,658	6,470	7,706		12,128	2,640		44,602
945							6,043	6,247	2,197						14,487
946						4,262	2,961								7,223
1025		489													489
1081															.
1840											526				526
66A		2,896	6,909	13,151	2,997			1,200	6,410	17,393	16,471	4,879	1,494		73,600
TB 9. Bl 1						958									958
9 Bl 2			598					1,804							2,402
9 Bl 3			4,493			4,290	45								8,828
Tbr. Permits								1,077	8,146	3,104		526			12,853
Adjust.		881		4,066											4,947
Totals MBM	4,267	14,663	28,130	24,424	29,069	25,379	42,749	49,042	57,606	54,241	28,493	43,089	40,885	37,262	479,299
=		Ex Faith.xls													401,152
=		Corroboration from sessional Annual Reports													370,861
		Volumes are in MBM.									Faithsort final.xls				479,299

Figure 34: Table 1, originally table 1A. From files with the Saskatchewan Archives Board and sessional reports, Department of the Interior.

406

This table 1A (see larger table at www.Silviba.com) contributed to the analysis of woodflows undertaken for Sturgeon Lake First Nation before its submission to Indian Specific Claims Commission. It sought compensation for timber stolen from this reserve between 1904 and 1909. It shows that PAL produced a reported volume 479 million board feet of lumber between 1905 and 1919. The timber agent in Prince Albert recorded all timber volume logged on each timber berth quarterly from duly sworn statements.

This table is combined from two sources. The first is the federal timber agent's annual reports. Second, these were cross-referenced with documentation in the NR5 timber berth files held in the Saskatchewan Archives Board in Saskatoon. Three further matters need addressing:

- Those 400 square miles of timber berths were turned over to PAL by the owners. I think that George Burn, our banker sat in Ottawa, received annual royalty cheques from PAL treasurer, Peter Pannabacher. By signing over control of his timber berths, he saved himself from having to undertake all the paperwork and rents connected with upholding the annual approvals to remove timber from Crown lands. PAL, in turn, got control of the timber. George would have received an annual payment of so much a thousand board feet.
- At start up in late December 1905, logs came from Telford's log yard in Prince Albert. Table 1 says that PAL produced some four plus million board feet of lumber by June 30, 1906.
- Some 70 percent of the content of the NR5 files concerned collecting annual licence fees and back dues for the timber logged. Perhaps, most importantly were the quarterly sworn returns for every active timber berth. Table 1A became the summary for all fourteen years of PAL active logging operations recorded through quarterly affidavits (category 1 timber volume). Technically, it did not shut down until 1921. However, after the 1919 forest fires, the Company floated the main mill over to Manitoba. The timber agent instructed PAL

to clean up the decked logs before cancelling all those timber berths.

Discrepancies

Sawmill owners had a few things up their sleeves when it came to avoiding timber dues and annual rents when recording timber volumes logged. They operated at least two sets of books: one for the government timber administrators and one for the shareholders. Any logger worth his salt would trespass and log free wood at the first opportunity – they still do.[115] With a stated lumber capacity of 50 million board feet yearly, you would think the market swings would even out after 15 years. This would mean a lumber production of 15 X 50 = 750 million board feet, or about 55% more than reported on paper. Also, being winter weather with snow matting on the trees, they never noticed the boundary lines – just the next juicy tree immediately ahead. And no spray paint and orange flagging tape in those days. The loggers started operations in December and finished when the ice began to melt in the spring.

A report prepared by the government of Saskatchewan, Department of Natural Resources in 1955 titled 'Saskatchewan's Forests' is quoted on page 6 as follows:

"Although there are only incomplete records on woods production, it was from this area (the commercial forest zone) that most of the spruce sawtimber came during the period 1900 to 1930. Prince Albert represented the 'economic capital' of this region and at the peak of the sawtimber industry it is stated that annually as much as 100,000,000 feet board measure of spruce was processed in that city. It is not inconceivable that Prince Albert Lumber Co. attached production levels approaching, or even exceeding 75 million board feet annually. The remainder would have come from a few smaller local mills scattered around the district."

These statements, combined with federal government audits of 1911 and 1920 pointed out that both Sturgeon Lake Lumber and PAL had acquired Indian timber. This would have been just a fraction of that removed in trespass. Only two audits in sixteen years.

408

From the Saskatchewan Archives Board NR 5 catalogue and associated files, also sessional reports, we can forecast a range of lumber volumes produced in Prince Albert between 1904 and 1919.

		Aggregated volumes (categories 1 and 2)							
		Annual lumber volume manuf., MBM, Category 1 and 2							
		by Prince Albert Lumber, 1905 to 1919							
					50 MMBM	% Mill	70 MMBM	% Mill	total
		Manuf vol	Category 2	Total vol.	Capacity	Capacity	Capacity	Capacity	Lath
		MBM	MBM	Manuf. MBM	MBM	Utilisation	MBM	Utilisation	Pcs.
1905/06	1906	4,267		4,267	6,000	71	6,000	71	4,663,400
1906/07	1907	14,663	15,000	29,663	50,000	59	70,000	42	6,742,600
1907/08	1908	28,130	15,000	43,130	50,000	86	70,000	62	6,896,000
1908/09	1909	24,424	5,000	29,424	50,000	59	70,000	42	6,154,800
1909/10	1910	29,069	5,000	34,069	50,000	68	70,000	49	6,067,800
1910/11	1911	25,379	1,000	26,379	50,000	53	70,000	38	5,400,374
1911/12	1912	42,749	1,439	44,188	50,000	88	70,000	63	8,723,028
1912/13	1913	49,042	682	49,724	50,000	99	70,000	71	9,808,400
1913/14	1914	57,606		57,606	50,000	115	70,000	82	11,521,200
1914/15	1915	54,241		54,241	50,000	108	70,000	77	10,848,200
1915/16	1916	28,493		28,493	50,000	57	70,000	41	5,698,600
1916/17	1917	43,089	1,139	44,228	50,000	88	70,000	63	8,907,106
1917/18	1918	40,885		40,885	50,000	82	70,000	58	8,177,000
1918/19	1919	37,362		37,362	50,000	75	70,000	53	7,472,400
Total MBM		479,399	44,260	523,659	656,000	80	916,000	57	107,080,908

Note: Lath volumes derived from a factor of 1 MBM = 200 pieces.

Ex Graphed PA lumber annual volumes.xls

Figure35: Table 2: forecasts of total PAL lumber volume between 1905 and 1919 at (a) 50 MMBM a year, or (b) 70 MMBM a year. Saskatchewan Archives Board, NR 5

Category 2 timber (unreported timber volume) was sourced from Sturgeon Lake reserve and from scaled wood removed by Sturgeon Lake Lumber from Little Red reserve, 106A. Category 2 volumes shown were a blend of the adjudicated timber surrender volume discovered through Department of Interior government audits. However, the Indians received no reimbursement in trust. For Sturgeon Lake reserve, a timber surrender volume of 41 million board feet was presumptively spread over five years. The Sturgeon Lake Indians received an adjudication of 41 million board feet valued at $4.4 million in 2001.

A fair settlement for the Little Red timber surrender claim is also needed. But this latter claim is still in dispute.

These photos were also taken by taken by Prince Albert photographer James on Sturgeon Lake in the winter of 1906-07 (Saskatchewan Archives Board, Regina.)

Figure36: (R – B1039). Note how the load has shifted forward. Saskatchewan Archives Board

Figure37: (R – A7091). Logs hauled on to the ice prior to decking, Sturgeon Lake. Saskatchewan Archives Board

Figure 38: (R – A3822). A procession of loaded sleds on its way to Sturgeon Lake from Camp 2. Saskatchewan Archives Board

Figure 28: (R – A1665). Two log trains arriving on Sturgeon Lake. Saskatchewan Archives Board

Note the camp building behind the first engine.

Figure 40: (R – A1622 – 3). Another train on Sturgeon Lake. Saskatchewan Archives Board

Note the two-horse sleigh and occupants. One thinks that this was the conveyance that brought photographer James and company up the Montreal Lake trail to Sturgeon Lake in the winter of 1907.

Figure 41: (R – A1665). Timber scalers with scale sticks recording logs by diameter and length as a basis for payment to contractors. Saskatchewan Archives Board

412

I believe that photographer James left Sturgeon Lake to take more photos up the Montreal Lake trail before heading back to Prince Albert or, perhaps, staying in camp 2 (on the north side of Sturgeon Lake and halfway along) for the night.

Loggers transferred all this wood onto the ice at Sturgeon Lake between December 1906 and March and April 1907. This amounted to 75 million board feet from Little Red and Sturgeon Lake sources. In fact, in October 1906 the Prince Albert Weekly Times newspaper reported that Sturgeon Lake Lumber Co was shutting down its camps for the winter. It had signed a logging contract for 17 million board feet with the British Columbia Construction Co.[116] Also, A.J. Bell let it be known to Indian Affairs that they could not start logging on IR 106A (Little Red) until they had finished up IR 101 adjacent in 1906-07(from archived correspondence on file).

LUMBER PRODUCTION, PRINCE ALBERT DISTRICT, DEPARTMENT OF INTERIOR.
Lumber manufactured from stationary mills, MMBM.

Year	Sanderson	Cowan	Shannon	Telford	Sturg Lk L	Other	P.A. Lbr	Total
1903/04	3.1	4.3	2.3	5.6		0	0	15.3
1904/05	2.6	4.6		5.4	3	0	0	15.6
			** mill burned down May 1905					
1905/06	4	2.9			4.6	3.5	4.3	19.2
1906/07 9 months	2.7	3.2			4	3.7	14.7	28.5
1907/08 15 months		4.5			1.25	6.8	28.1	50.2
1908/09							24.4	
1909/10						8.6	29.1	41.4
1910/11							25.4	25.4
1911/12							42.7	42.7
1912/13							49	49
1913/14							57.6	57.6
1914/15							54.2	54.2
1915/16							28.5	28.5
1916/17							43.1	43.1
1917/18							40.1	40.1
1918/19							37.3	37.3
							548	million board feet

Note:

Information supplied by Alexander Dietz from sessional papers, U. of Saskatchewan.
"Other" sawmill volumes are not tributary to P.A. Lumber wood supply area.
Source: Sessional reports, dept of Interior cross-referenced with TB annual returns submitted.
Silviba   sturg\anrpt.xls

Figure42: Table 3: All reported lumber production ex Prince Albert from 1906 to 1919. Saskatchewan Archives Board catalogue NR 5

413

Figure43: (R – D1054 – 1). Preparing the skidway into the next block of timber. Saskatchewan Archives Board

One source points out the loggers spaced skidways 200 feet apart.

Figure44: (R – A8615 – 4). Wow! Saskatchewan Archives Board

Figure45: (R – A1621 – 3). Log decks waiting for spring break up – Sturgeon Lake. Saskatchewan Archives Board

Figure46: (R – A8615). They logged the hills surrounding Sturgeon Lake clean by the spring of 1907. Saskatchewan Archives Board

Figure47: (R – A1621 – 3). Another train of logs arriving on Sturgeon Lake. Saskatchewan Archives Board

Figure 48: (R – B1702). In 1910, log chute with Little Red River running south. Saskatchewan Archives Board

Albert Mattes brought in a river drive engineer to improve river flows on the Little Red. They built this dam just below the confluence with Bell Lake. Note the sluice gates – balance of probabilities. (There is also a photo in the archives of the dam across the River.) Peter Pannabacher, the PAL accountant, audited log losses from each active timber berth after the 1910 spring river drive. The Company immediately brought in a river-drive engineer to reduce log losses to the mill.

Figure49: (R – A1622 – 3). This could be the Whitestar bridge on the Prince Albert to Montreal Lake

Ω

PAL's timber supply burned up in 1919. Massive fires spread right across the Parklands caused by settler fires. The Company loaded all of its machinery on to barges, then rebuilt at The Pas, and became The Pas Lumber Co. In 1956, the CCF forestry minister for Saskatchewan terminated its remaining timber limits on Saskatchewan's east side. It re-opened in Prince George as The Pas Lumber Co. shortly after. Even today, you can see the Winton logo in red (a 'W' superimposed on a circular saw) on the lumber wrapper of every lift of lumber originating from that mill, either at your local building supply outlet or in transit by rail or truck.

Prince Albert faded back into the woodwork after that, or what was left of it.

Ω

417

PART FOUR

Discussions, conclusions and recommendations

From my long-term exposure to forestry, I gather recommended frameworks with respect to:

- Forest administration in both Saskatchewan and British Columbia

- Indigenous peoples in Saskatchewan

Ω

The following are my ideas, opinions, conclusions and recommendations for the better direction of the future for public forest administration. These include optimisation of the *Candian* contribution in a parallel universe. Canada will thus have three universes, all distinct but making up the whole cloth: Canadian French, *Candian* association and the rest of us (Can-mongrel).

Chapter 33:

New forest administration considerations for Saskatchewan and BC

The following are my ideas and opinions for improved performance of public forest administration.

Forests and other land vegetation currently remove up to 30 percent of human carbon dioxide emissions from the atmosphere during photosynthesis. If the rate of absorption was to slow down, the rate of global warming would speed up in return. (Nasa.gov/earthgrightnow). Although, forests in the northern hemisphere contribute far less removal than forests in tropical ecosystems. The provincial harvest volume schedule becomes a proxy measure of carbon sequestration – add in, or subtract, humus, leaf litter and twigs. Not to forget water. Unfortunately, the forest fires of 2015 have released more carbon than is captured. Although this is just an educated guess.

Today's public forest administration lacks accountability and measurable objectives. Turf, politics, uncertain budgets and absence of good data reduce the effectiveness of management and desirable outcomes. Also, variable budgets defer reaching objectives – look at BC's replanting backlog. Does a cabinet minister lose his job because of failure to achieve targeted objectives? Hardly. Because there are no penalties for missing these. The real punishment ends up, not in the forest industry because of shrinkage in the harvest volume schedule, but with the public through job losses and a shrinking economy.

The days of long-running appointments such as Honourable Ray Williston's tenure as minister responsible for the BC Forest Service, have gone. He was an excellent choice. Forest management is a long-term commitment. In my day, in Saskatchewan, forestry was part of the training ground for new ministers with limited power at the cabinet table. Probably a deliberate policy, given the forestry Crowns ruled the roost.

We need an approach that combines the management of economic outcome with those of the Environment. The measurable objectives include jobs, recreation and cashflows. All within a stable ecosystem, including climatic vectors. Our forests are a business asset. The least we can do is maintain the asset value, or grow it. The public forest estate should include a healthy forest economy, and growing assets that incorporate carbon sequestration and cap-and-trade choices. Similarly, Canadian Indian (*Candian*) traditional land managers could consider a separate, yet parallel forest policy that meets the wishes of First Peoples. Remember that diversification is the key to long-term family survival.

This engine needs to overcome deficiencies obvious in the existing public administration systems. These aggravate barriers to growth and introduce inequalities.

Small vs large forest operators

Generic definitions

Large forest operators integrate pulpmills with sawmills. Alternatively, they ship wood chips elsewhere. This category also includes board plants (OSB) and plywood. The investments are large, $1 billion plus. They have access to the province's government through Executive Council. They give generously to political campaign funds. Their timber supplies are assured through long-term forest tenures, renewable. Forest administration is conducted through a contract between the government and the forest operator.

They require large pools of capital to integrate their operations to maximize the return on investment through income trusts. Any company, division or subsidiary not making 26 percent is sold or dismantled, as seen with Weyerhaeuser Canada – now just a 'has been' except for the odd branch still kept alive. In my day, a company with a well-managed forest supply made 7 percent on investment. The forest base was well regarded by the owners. Company foresters conducted research projects aimed at improving forest sustainability and growth. Today, all that energy is transferred to the bottom line. Shareholders have no stake or commitment to the forest estate. Neither does government, the ultimate owner – for all citizens. Today, pulpmill cashflow relies on the sale of excess electric power through cogeneration.

Income trusts have denuded the landscape when it comes to forest replacement and programs designed to enhance growth and yield. You could conclude that 26 percent minus 7 percent = 19 percent of the bottom line is perhaps the cost of sustainable forest management. Instead, this residual goes to the shareholders who have no obligations to renew the forests of Saskatchewan or BC. Foresters, either in government or the forest industry, become pawns in the game.

Intermediate forest operators are small operators on a growth spurt. The old guard died, and was replaced with ownership full of energy and ambition. It invests, modernizes, and diversifies. It is canny and has an 'in' with key individuals within government. Marketing skills are top drawer. The ownership retains its roots, and it is 'bloody minded' at times. Personalities are in play. It hates bureaucratic delays and obstructions. It is 'our way or the highway.' However, this class of ownership is still independent. It is not saddled with decision-making constraints outside of its own scope. This promotes fast decisions, innovation and flexibility. It will deal with any legitimate constituency to get the job done. It tends to be 'cheap' with its contractors, having been there on the way up. It avoids taxing long-term forest management duties. Although it will provide reforestation dollars and forest management fees when needed to. Intermediate operators rely on a renewable Crown timber supply allocated ten years at a time.

Small forest operators extend the provincial allowable cut. They utilise smaller forest parcels and can exist within a niche market. They evolve independently and are loathe to associate with other like-minded loners. The survivors have honed their marketing skills and now have purchaser connections. They have built a solid home life thanks to wives and families with unlimited patience and endurance, however thin, as business absorbs most free time. They resent intrusions on their operations from government agencies with which they live reluctantly. The need for survival is acute. Only one small sawmill operator I worked with always got money back from the annual government audit of his sawmill records. Some four days before the annual audit, he disappeared into his home office. After he emerged, he had a perfect set of sawmill records. They lived uncertainly since government limited their access to Crown timber to annual permits. These are not necessarily renewable. Thus, they struggle eternally with finances.

In my Silviba life I saw one experiment with small operator timber supply. Newton Logging's annual timber allocation was in the heart of Weyerhaeuser's operating area east of Caribou Creek and north of Smeaton. Garry Newton arranged with Weyerhaeuser to deliver logs direct to his sawmill. Weyerhaeuser took care of the paperwork. He did have trouble preserving log quality – Weyerhaeuser also loved sawlogs – but he had no administrative overhead. I believe Weyerhaeuser added a small handling fee to its charge for the logs delivered. A small sawmill could go bankrupt if the log supply became incompatible with the mill design – not enough lumber recovery for the dollars paid. Later, Weyerhaeuser's bush foreman changed. The sawlog sizes shrank and Garry ended the arrangement.

Here, on the Sunshine Coast, the small sawmills rely on 'handouts' from the resident forest monopolies. Most have shut down because of misgivings attached to log quality and volume available. The few that remain survive vicariously on timber from private land, with one exception. His sawlogs are purchased elsewhere from salvaged timber .

Ω

Forest policy in Saskatchewan was lacking any guidance for civil servants in managing timber supply planning for small industry. The power at the cabinet table ensured the large industry got the cake in both Saskatchewan and British Columbia. There used to be crumbs for small operators. Not anymore. This created huge stress for key individuals in Forestry Branch, as well as for individual operators. The guidelines conflicted. The winners for the timber supply lottery insured themselves forever if they held a large-scale forest management contract with renewable terms. Or a few years back when every outfit with clout was a Crown corporation. The Crown sawmills in Big River and Carrot River opened. The Saskply plant at Hudson Bay started as a joint venture, but eventually stayed in government hands until the coming of Weyerhaeuser in 1986.

Figure 50: http://saskhistoryonline.ca/explore. Brings back memories.

Small operators, mutations born after the Saskatchewan Timber Board era, existed in communities across the Parkland region. A DREE[117] forest products directory in 1979 listed 249 or so small mills clustered in communities across the forest belt. They worked the Prince Albert . no. 2 portable mill equipped with a 50-inch diameter circular saw, 3/8 saw kerf.[118]. They powered the sawmill with either a tractor and power take off, or with a diesel engine. With the onset of forestry Crowns, and later, Weyerhaeuser's entry into Saskatchewan, and a hostile Forestry Branch, the number of small mills shrank to less than five in the province.

All joined in the competition for spruce timber. The Saskatchewan government amended the Forest Regulations. After 1966, small operators could no longer permit any spruce timber over eight inches in diameter. Larger trees were reserved for the Crowns. Luckily, large forest fires provided for small operator salvage cuts

425

from time to time. The big guys stayed away from fire kill if they could. Wood chips containing flecks of char were taboo in the pulpmill. To irritate matters further, the Crown pulpmill converted large spruce sawtimber to pulp. Weyerhaeuser followed with this practice as well – despite an integrated forest economy by 1986. Timber regulation should have directed any log smaller than six inches to the pulpmill. The sawmills would utilise the balance. Even the plywood plant retooled to peel smaller diameter logs. Depending on markets, Weyerhaeuser utilised timber down to three inches top diameter at times. I was a proud Saskatchewan forester in those days. Weyerhaeuser at Big River sold all of its oversize logs to small operators. The ring debarkers at the sawmill were too small to take a white spruce sawlog twenty-two inches in diameter. Today on the Sunshine Coast, I see nothing going down the highway under seven inches. It leaves me sick at heart that during my thirty-three year absence from the Coast, timber utilisation is worse, not better. Why? Because we export so much of it in log form. Loggers leave the rest in the woods. Shame.

I have to tell a story. Gene Kimbley had a job with the old Department of Northern Saskatchewan. He supervised a few sawmills throughout the north. The sawmill at Deschambault started out as a DNS mill. The Department also contracted out lumber planing to Kosowan Bros at Garrick (between Choiceland and Love on Hwy 55). One afternoon, Geno went to the mill to take inventory of all dressed lumber in the yard. DNS had just folded up and was doing an inventory of all possessions. Lo and behold, the yard was empty – someone had sold every available stick of dressed lumber. Geno never forgave the Kosowans, and muttered about his trip to Garrick for years. None of this was ever proven though. The dominant theory at the time was the Kosowans sold the lumber to recoup costs incurred. Later, Greg Kosowan logged for Barry DeVocht at Provincial Forest Products Ltd. Remember, he was logging at Weyakwin when that plow wind came through and white spruce sawtimber fell right beside the bunkhouse trailer. Eventually, the Kosowans moved to Alberta.

In the largest market downturn since the dirty thirties, the Crown pulpmill in 1982 reduced to one shift because of a reliance on market pulp. Extended shutdowns were the norm. These also applied to the Crown sawmills. Small sawmills continued to process custom orders for customers in Winnipeg, Toronto and Montreal. They struggled, but preserved payroll and kept their suppliers with product. The Weyerhaeuser building supply centre in Winnipeg bought dimension lumber and large diameter beams and stringers from Saskatchewan small sawmills. Remember that Provincial Forest Products Ltd in Prince Albert found a big market for poplar ties used to support pipe during pipeline construction. The fencepost treatment plants continued shipping product throughout western Canada and into the States.

Ω

There are four advantages to preserving a balanced timber supply between small and large forest industries in Saskatchewan:

*Small operators have smaller overheads. This means they can log smaller patches of timber. A large Weyerhaeuser operator with his mechanical logging equipment needed twenty ha each shift ahead of the feller-buncher. From a large operator perspective, any forest patches under 20 ha were uneconomic. Thus, the provincial timber supply shrinks by twenty-five percent – about 1,700 direct jobs throughout scattered communities. A small operator unit makes use of this. We talk 'jobs' in government, but no one in government knew that a small sawmill supports one full-time direct job for every 525 m3 consumed. In Weyerhaeuser's case, this value was 2,100 m3. Thus, small operators create almost four times the job numbers per cubic metre than Weyerhaeuser.

*Small operators increase the allowable cut by salvaging windfall and fire-kill timber. Add this to leftover bluffs of good sawlog material.

*Large industries concentrate wealth, whereas small operators existed in as many as sixteen communities scattered across the Parklands. They dispersed wealth. This is important because the history of some small sawmills lies in their integration with farming. They spread cashflow more evenly across the seasons of the year. Lorne Colvin, who farmed on Highway 55 between Nipawin and Whitefox, was a marvelous example of a farm-forestry economy. This region was also the home territory of the White Spruce Lumbermen's Association. The province loses another fifteen percent of the provincial allowable cut because of losses through fire, insects, disease and blowdown. So, theoretically, we add another 500,000 cubic metres of annual allowable cut, meaning another 3,000 rural forestry jobs. I can just hear my industry colleagues, government foresters, plus the Ministry of the Economy, in total denial. When allowed to, small operators do salvage the blowdown and fire-killed timber.

The same is true for British Columbia. Vested interests want no part of it. There is one problem though. Often, a small operator will have to build road. However, his financial depth precludes this capacity. The solution is easy. The Crown advances funding, paid back per cubic metre of logs removed. Don't let the Crown build the road – it will cost twice as much. This is possibly a legitimate use of a Crown corporation – administration of a provincial logging road loan agency. In Saskatchewan, logging roads were built by Saskatchewan Highways. Then came Weyerhaeuser. At first, the government continued to build 32 km a year. Soon, logging road construction reverted to industry because of NAFTA's perception that government was subsidising the forest industry.

*Small operators supply fencepost peddlers with rough lumber, corral fencing, barn poles and rails. The lumber on those ten-ton trucks heading south from Prince Albert on a Sunday night ended in Shaunavon, Mankota and Estevan, Saskatchewan, as well as farmyards on the way. Aallcann Wood Suppliers strapped down farm lumber on top of the fencepost bundles heading into the States, as did Ron Lehner of Lehner Wood Preservers in Prince Albert. Ron was his own manager, truck driver and peddler. I enjoyed having breakfast with him at Macdonalds on a Sunday morning. It was difficult for anyone to find enough rough lumber to complete the orders. Even bundles of second-cut slabs.[119]

Small operator timber utilisation is primitive. One cubic foot of logs = six board feet of lumber, as opposed to a large, integrated operation that produces nine board feet from the same cubic foot of raw log. Add to this, the fact the little guys leave behind slab piles, while the big guys chip it all up, and everything, including the squeak, is processed. Timber rules should include the addition of a resaw and the direction of subsequent waste to an electrical cogeneration plant. The mill then value-adds those slabs into one-inch lumber. A good percentage of clears comes off the resaw[120]. And government power corporations receive electricity.

Value-adding

Talking about 'value-adding', this was the Forestry Branch mantra in the latter part of the 1980s. It was the total solution to every dilemma imaginable. Government bought into it big time. But no one defined its meaning. It was like religion. Some examples of value-adding are:

- Converting birch into furniture. I surveyed 17 million board feet of mature birch on the Kelsey Trail east of Nipawin for setting up a small sawmill to produce furniture grade lumber. This was a serious dream. The economics were negative.

- Here on the Sunshine Coast one cabinetmaker uses alder to make the panels and framing for kitchens. Remarkably free of knots. Beautiful products. Another guy manufactures high-priced dining room chairs from maple and alder.

- A home repair business installs birch flooring using local materials.

- A few carvers all over create bowls and ornaments from wood burls and other wood anomalies.

- Another connection here on the Sunshine Coast saws custom lumber sizes with a portable sawmill from cedar scrounged in local log sorting grounds. Five other local sawmills of the small operator category closed due to shortage of cedar. It is too high-priced and is shipped in log form off the Sunshine Coast. They had no formal quota.

- Pellets from waste wood and electric cogeneration from sawmill leftovers are value-added. There is ample room here on the Sunshine Coast and in Prince Albert to get into wood pellets. Mainly for industries burning coal. The addition of wood pellets cuts back on toxic gas emissions.

- Barry De Vocht's production of poplar ties for laying pipelines from unmerchantable poplar is good example.

- Garry Newton at his Cariboo Creek sawmill north of Smeaton installed a system that enabled him to make tongue and groove jack pine lumber for installation on rec room walls. It was decorative. The blue stain evident in fire-killed pine added to the value. He developed a market niche for Cariboo pine rec room panelling.

- Conversion of first- and second-cut slabs to marketable products. The farmers take some. Others need remanufacturing into one-inch lumber.

Saskatchewan pioneers made roof and wall shingles from poplar and jack pine. And log homes. They floored grain bins with cottonwood lumber. Most of the foregoing is partly classed as 'cottage industry', thus has limited economic clout. However, it shares one significant advantage – it dispensed wealth and was community-based. Remember: a balance between large forest industry and community-based forestry is the ingredient in a healthy provincial forest economy. They need each other.

Ω

Another fall-down for the administration of small operators from the government perspective is that small operators use up much government staff time. The solution was to permit them just enough timber to cover the size of a postage stamp, and just for one year.

When I came to Saskatchewan in 1981, resource officers in Forest Operations were based at Meadow Lake, Prince Albert and Hudson Bay. They were supervised by the operations forester headquartered in Prince Albert. It was a collaborative administration arrangement that solved issues on the ground. The resource officer could always count on a cup of coffee, or lunch at the sawmill, or in the bush at the logger's camp. The relationship was symbiotic.

430

Today, conservation officers, complete with firearms, are too ready to treat the small operator as if he was a fugitive from justice. As mentioned earlier, in Prince Albert, six conservation officers were wedged into a ground floor office and shared two pickups among them. There has to be a better way to administer well-designed forest regulations in Saskatchewan. However, before this is countenanced, someone has to produce the well-designed legislation.

In my experience, a civil servant uses the rules empathetically in dealing with a client. But a bureaucrat hides behind the regulations, feels threatened if a decision is needed, and puts his personal needs ahead of the client's, mainly pension credits. Mr. Don Bernier was a Forestry Branch employee in the days prior to digital data. His job was to produce forest inventory maps ordered by 'publics' such as Silviba. Without them, I could not service my clients. Don never missed a target time for map delivery. I nominate him as the quintessential civil servant in Forestry Branch in the pre-digital epoch. In today's world of digital forest inventory information, his successor must be David Lindenas, manager of forest inventory for the Saskatchewan government. Saskatchewan Environment also contained an outstanding civil servant. His job was to set out the rules for environmental impact assessments. He followed up with nurture and advice. The result was an improved EIA and a successful outcome. I do not remember his name though.

Once, a small operator asked that I go with him to seek a cutting permit from a forest operations official in Prince Albert. We showed up at the stated time of eleven o'clock, and found the guy had gone home. So, we made an appointment for the next day at one pm. He was there, and the sweat from his armpits was staining his shirt. He was agitated and looked stressed out. He would love to have said 'no', but he knew the heat would intensify if he turned down the small operator. In fairness to our official, Weyerhaeuser had its thumb on him big time – he was not to give away 'its' timber to small operators.

Today in the western provinces, field staff has shrunk significantly. A deregulated arrangement exists. The large operator is also his own policeman. Last, but not least, is the place of fencepost operators in the Saskatchewan forest economy, especially around Prince Albert and Glaslyn. The pulpmill woods staff blocked access

431

to jackpine fencepost timber whenever possible. I remember one prime stand of fencepost timber on dry ground at the Besnard junction with Highway 2 north. It was on Aallcann's cutplan. When Aallcann's cutters moved in, the pulpmill had logged that timber. No friends in Forestry Branch or a conservation officer to serve as intermediaries. One supposes the conservation officer was without a pickup that day.

I cannot speak to the progress made with Saskatchewan forest administration since the arrival of *Sakaw Askiy* Forest Management Inc., replacing Weyerhaeuser's role in Saskatchewan forestry. Alphonse Bird hired me to help him with representing Montreal Lake Cree Nation. We were in the boardroom at the Forestry Centre in Prince Albert in the fall of 2010. I realise now that Montreal Lake Indians were given 150,000 cubic metres of the annual cut. A large timber share instead. How many jobs created for this Cree Nation? As this is written, I suspect that most of it burned in 2015 forest fires.

I must say, though, that *Sakaw Askiy* is a positive government initiative. It has great potential to enhance socioeconomic gain over the midterm. However, it is a risky venture. Companies have staked out their turf without realizing that a forest fire could wipe one of them out, just as with Prince Albert Lumber Co. in 1919. What do they do then? Shut her down and declare bankruptcy, or rely on another company to move over and let them into a substitute patch of timber while using the other guy's road system? This will not happen. Will Montreal Lake take its allocation of timber and rent it to others for 50 cents a cubic metre? No job creation for Montreal Lake Cree Nation. Just more casino funds. On the other hand, if MLCN residents pick up some of those bush jobs, perhaps this becomes a more realistic solution. I reference Tolko Industries' integration of band members into their logging operations – both at Sturgeon Landing and Deschambault.

The Tolko model with the spruce timber salvage on Sturgeon Landing IR 184F was the best short-term model discovered by me. Tolko hired a trainer to teach nine band members to operate equipment. Each took home $1,800 from every payday. It is the best choice, it seems, at least for now. See an account of that logging operation in Chapter 28. I know the folks at Carrier Forest Products in

Saskatchewan will work out a mutually useful arrangement with the renters of timber.

Do not forget that Montreal Lake did operate a contract logging operation that made money. Then Chief emptied the bank account and left over $1 million in logging equipment in the boneyard behind the development corporation shop. As a forester, I find this heartbreaking. Of all homeland communities north of Prince Albert, it has the best chance of combining its traditional values with today's need for community-based payroll. However, a succession of chiefs and councils put themselves first, while ignoring the needs of their people. I saw my first smashed up hockey rink at Montreal Lake. And, where did that $1.25 million dollars the Minister of Economy gave to Montreal Lake Cree Nation end up?

Provincial timber supply should accommodate both community-based small operators and larger industries. The former disperses wealth locally. The latter concentrates wealth within a region or province.

Alternate forest administration models

During my career, I have detected two forest administration/implementation models that need exploring by someone who could take this on as a mission.

One: Forestry Commission Scotland model

After returning from my last stint in Russia, I visited Tony Baumgartner in his Ministry of the Economy enclave in Regina. My visit was to sow the seeds of a study into the organisation, actions and operating procedures within the Forestry Commission Scotland. I should have gone to the minister. I had worked with Robin Cutler, the former chief of this agency during my Russian work. I recommended that Saskatchewan should study the model. Perhaps consider it for adoption.

"Forestry Commission Scotland advises and implements forestry policy to protect and expand Scotland's forests and to increase their value to society and the environment."[121]

Robin Cutler volunteered to come to Saskatchewan and provide an overview of that agency; after all, he used to run it. The attraction for me as a forester is in the holistic nature of the organisation and its objectives. All sectors relying on access to the forestry landscape are represented. Although separate, they come together under one roof.

For Scotland:

From public forest lands, 2.78 million m3 in the 2002-06 period, rising to 4.432 million m3 in the 1917-2021 period. A growth rate of 160%. From public forest domain.

From private forest lands (tree farms), 3.415 million m3 in the 2003-06 period, rising to 5.713 million m3 in the 1917-21 period.

I wonder if Saskatchewan has ever set a forest growth target. Something to consider?

Ω

"Economic Analysis of the Contribution of the Forest Estate Managed by

Forestry Commission Scotland (FCS)

Final report for Forestry Commission Scotland, 2004

2.1.3 Timber output and the processing sector

Production forecasts of softwood availability give a public estate output of 2,782,000 m3 per year in the 2002- 2006 period rising to 4,432 '000 m3 in 2017-2021 (Forestry Commission, 2004a). This compares with 3,415 '000 m3 rising to 5,713 '000 m3 for the private sector. FCS (2003) states that the estate produces around 60% of the wood production in Scotland. It is thus the major supplier of domestic timber for processing and has been a predictable source of supply under long-term contracts for processors. Eiser and Roberts (2002) estimated the employment multipliers from coniferous planting/maintenance and harvesting as 0.49 and 11.78 FTE per 1000 ha.

Thus, reducing the area of coniferous forestry by 1,000 ha would reduce knock-on Employment by around 12.3 FTE jobs." – Source, Forestry Commission Scotland, website.

Ω

I asked FC Scotland for an opinion on the probability that we could run into a problem with NAFTA due to a perceived subsidisation from running a provincial forest administration system using a 'commission 'model:

Email: May 6, 2015

Further to your request I think the best thing I can do is to give you a brief outline of our approach to marketing and selling timber. Forestry is now a devolved activity in GB and therefore the detail below refers to Forestry Commission Scotland alone. Forestry Commission Scotland manages the National Forest Estate (NFE) on behalf of Scottish Government.

The total production/sales from the NFE is 3.2m3obs/annum on a sustained basis, representing 40% of the total production from Scotland's forests.(N.B. Higher in FY15/16 due to Plant Health harvesting). So whilst the Private Sector produces 60% of the total, FCS is the largest single player.

All of our sales are marketed/entered in to competitively.

Approximately 45% of our timber is harvested by our own resources and either sold at forest "roadside" or delivered to the processor. The remaining 55% is sold "standing" with the purchaser responsible for harvesting and delivery to the end user.

We market timber (both roadside, delivered & standing) via a mixture of "open-market" short-term sales of less than 1 year's duration (approx. 33% of total) and Long Term Contracts LTCs, typically of 10 years duration. (approx. 66% of Total).

The open market sales are conducted via our open market electronic tendering system. We conduct 4 tenders per year, and lots are awarded primarily on price alone. The following link will take you to our e-sales web pages.

http://www.forestry.gov.uk/forestry/HCOU-4U4JGU2

Long Term Contracts are also marketed competitively, but the award process is a combination of price and a written submission. Bids are assessed using a scoring matrix against the aims and objectives of the Scottish Government's Forest Strategy in terms of their business, economic, social and environmental values. Once awarded the price will be adjusted usually twice annually(roadside sales) or on a coupe by coupe basis (standing), with the price being linked back to the original bid price.

I attach a summary of our sales plan for Forest Year April 15 to March 16, which provides the detailed split outlined above.

In terms of financing our main sources of income are timber, minerals and renewables (wind & Hydro); income from these finances our main forest management activities. In addition to this we receive funding from the Scottish Government (currently £21.7m/year) to support public benefits on the National Forest Estate e.g. recreation, environment, heritage, communities etc.

I trust the above provides you with the necessary insight you require.

Best Regards

Mike Green

Harvesting & Marketing Officer

Forest Enterprise Scotland

South Scotland Management Office

55/57 Moffat Road

Dumfries

DG1 1NP

mike.green@forestry.gsi.gov.uk

+44 (0) 1387 272440 (Switchboard)

+44 (0) 1698 222445 (Direct)

+44 (0) 7771 805229 (Mobile)

www.forestry.gov.uk/scotland

www.facebook.com/enjoyscotlandsforests

Forest Enterprise is an agency of Forestry Commission Scotland charged with managing the National Forest Estate.

From: Green, Mike

Sent: 27 April 2015 17:34

To: silviba@eastlink.ca

Cc: Green, Mike

Subject: Enquiry"

Ω

Now I know why 'BC Timber Sales' was born. This arrangement mimics FC Scotland. It sells Crown timber organised into logging blocks through tenders from private loggers. The object is to keep NAFTA happy. These folk, in turn, send it on to the manufacturer with the best return for the monies expended. Tragically, BC exports significant jobs overseas through round log exporting rather than processing the logs at home. At least Saskatchewan processes some of its wheat into pasta.

A. Proposed Forestry Commission Saskatchewan (FCSask)

Objectives
The mission is to enhance ecosystem stability within the Saskatchewan forest landbase, in the face of climate change, by both expanding the harvest volume schedule and increasing the intensity of human interaction. This includes both economic and human components, and optimises carbon capture/cap and trade.

Measurable objectives need setting for:

- Forest renewal targets
- Recreational use growth targets
- Wildlife management targets
- Fisheries management targets
- Business development targets
- Enterprise revenue targets
- Logging waste assessments
- Forest ecosystem biomass and climate change impacts research
- Carbon capture targets
- Information Technology improvement targets

Organisation (FCSask)
The composition of the board of directors simulates properties found with Forestry Commission Scotland board composition. The chart displays the essence of intent.

It goes without saying that any presentation laid out in this chapter is subject to rigorous review and analysis.

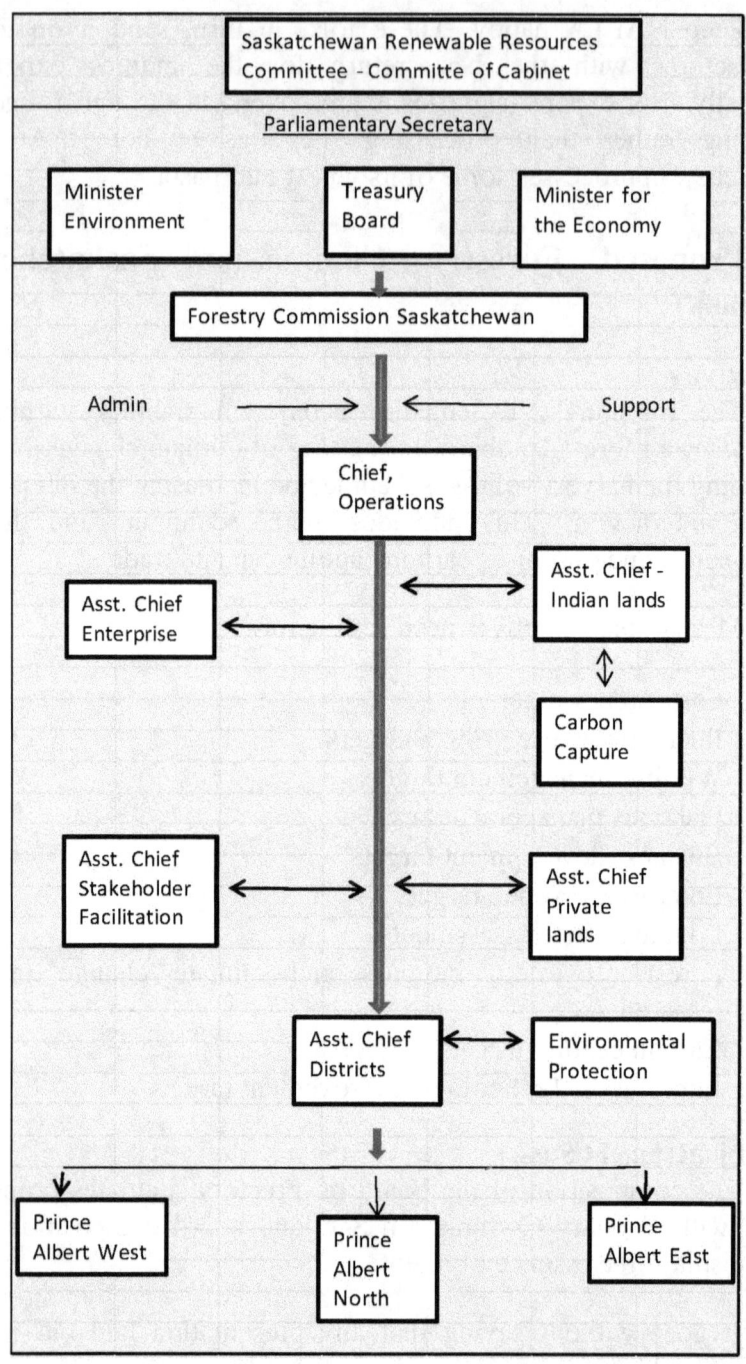

Figure 51: Pro forma organisational structure for Forestry Commission Saskatchewan

Forestry Commission Saskatchewan (FCSask) emulates the Forestry Commission Scotland template. FCSask is at arm's length through a funding agreement with the province. That entity, in turn, will treat with FCSask to transfer a portion of funds collected from operations to provincial revenue. The legislation for the startup of the Commission would include the essence of the Saskatchewan Forest Resources Management policies.

*Asst. Chief, Enterprises is responsible for all forest management of the Crown estate including timber felling, forest renewal, access roads, and forest administration. He/she reports to Chief, Operations and delivers programming through Asst. Chief, Districts.

*Asst. Chief, Indian Lands performs the same function as Asst. Chief Enterprises. Staffing emanates from the homeland communities. Also, this position administers all carbon capture programming on Indian, Crown and private lands. The position will be staffed from *First Candian* sources.

*Asst. Chief, Stakeholder Facilitation handles all interface communications with impacted parties – directly or indirectly. Also administers fishing and hunting. This position also presents all Enterprise planning before logging starts. The position includes the preparation of web-based reporting on all facets of FCSask, most particularly the state of carbon capture and progress with the growth of provincial annual allowable cut, both hardwoods and softwoods.

*Asst.Chief, Environmental Protection delivers forest fire management programs, and insect and disease management programs. He/she works through the districts. Also, liaison with Asst. Chief Indian Lands produces staffing for the forest fire management service.

*Asst. Chief, Private Lands delivers programming aimed at increasing carbon capture, business creation and growth in allowable cut levels. This position liaises with Sask agriculture programming aimed at fruit tree and berry management, amongst other initiatives.

*Asst. Chief, Districts as deputy Chief of Operations. All program delivery funnels through his office in Prince Albert.

Organisationally, many programs may need this position for coordination purposes rather than for direct delivery. The Commission will use the Forestry Centre in Prince Albert as its HQ. Chief of Operations will work out of Saskatoon.

*District superintendent for each of Hudson Bay, Prince Albert and Meadow Lake.

FCSask incorporates actions based on an action plan. It produces an annual revenue-cost forecast by months for all forest districts and programming. All parks north of Saskatoon will be including Cypress hills. These become the responsibility of FCSask.

Exciting, eh?

The birth of FCSask could become a major entity in a future election within Saskatchewan. In the reorganization to follow, staff will have an opportunity to apply to FCSask for the necessary positions. All audit services in the field will be contracted out. The application of a VAT percentage will apply to sales for all roundwood, slabs, chips, and other minor forest products. Other contract services include Support services, also the maintenance of an ongoing information base. *Sakaw Askiy* will have a subcontract to manage the lands assigned to it by the current arrangement.

Operators will deliver all waste wood to a cogeneration plant. Currently, this is dormant in Prince Albert. However, its failure to operate causes severe restriction to waste wood utilisation in Saskatchewan. Wood pellet manufacturing is also a genuine option from waste wood. Preferably, the Minister for the Economy should seek an expression of interest for wood waste conversion. Perhaps Montreal Lake Cree Nation and/or Prince Albert Grand Council should investigate the options. However, until political will solidifies, any further treatment of this topic is premature.

The bottom line is community economic health improvement in smaller communities across the Parklands with ecosystem and economic health gains.

A forestry commission has excellent advantages as the engine for managing and executing a forest operations action plan:

It answers to a board of directors.

It is decentralized operationally, but answers to a minister of state (undersecretary) within cabinet.

It is geared to increasing the volume harvest schedule, opportunities for new forestry investments and the enhancement of carbon capture.

It coordinates ancilliary uses such as hunting, fishing and recreation, and Parks. Also, water.

Organisationally, it is set up to manage and operate all things 'forestry'. It is given a mandate through legislation. It combines function and process to achieve its business plan outcomes. This includes northern Parks.

The composition of the board of directors simulates properties found with Forestry Commission Scotland board composition. The chart displays the essence of intent.

Forestry Commission Saskatchewan (FCSask) emulates the Forestry Commission Scotland template. FCSask is at arm's length through a funding agreement with the province. That entity, in turn, will treat with FCSask to transfer a portion of funds collected from operations to provincial revenue. The legislation for the startup of the Commission would include the essence of the Saskatchewan Forest Resources Management Act. However, program delivery is a blend of private (directly) and public enterprise (indirectly).

Ω

This is a ludicrously scant outline of the workings of Forestry Commission Saskatchewan (FCSask). But it serves to point us in a direction. When blended in with the Prince Albert Model Forest model, we may have a renewable resources management structure that puts Saskatchewan in the forefront of modern forest administrations within Canada.

After laying out the basic ingredients for FCSask, I note a structural resemblance in the way our forest resource is organised today. But it is split between three ministries – Economy, Environment and Culture-Parks-Sport. Also, it is process-driven. This weakens the intent to manage an asset. FCSask is a blend of corporate vs public enterprise. I am not certain if Parks administration stays *in situ* or evolves through FCSask.

FCSask could become a major entity in a future election within Saskatchewan. In the reorganization to follow, staff will have an opportunity to apply to FCSask for the necessary positions. All audit services in the field will be contracted out; the application of a VAT percentage will apply to sales for all roundwood, slabs, chips, and other minor forest products. Other contract services include Support services, and the maintenance of an ongoing information base. *Sakaw Askiy* will have a contract to manage the lands assigned to it or through modification of the current arrangement. There should be a timber allocation set aside for communities, say for operators requiring less than 1,500 cubic metres annually, or at the discretion of FCSask.

Operators will deliver all waste wood to a cogeneration plant. Currently, this is dormant in Prince Albert. However, its failure to operate causes severe restriction to waste wood utilisation in Saskatchewan. Wood pellet manufacturing is also a genuine option from waste wood. Preferably the Minister for the Economy should seek an expression of interest for wood waste conversion. Perhaps Montreal Lake Cree Nation and/or Prince Albert Grand Council should investigate the options. However, until political will solidifies, any further treatment of this topic is premature.

Two: Prince Albert Model Forest model (PAMF)
There are a number of ingredients involved in developing forest policy:

- Connected to the forest landbase. The forest resource and 'those that live in her' require a sustainable economy and environmentally stable ecosystem within which to flourish.
- The Forest Estate exists in a multi user envelope.
- Forest resource management accommodates climate change. Climate change expertise is required in the construction of a model that incorporates optimum carbon capture, socioeconomic return, and benefits to the citizens who own the Forest Estate.

- Carbon capture as part of the ecological provincial inventory along with timber and non-tangibles

This is a really broad set of objectives. The PAMF did not get beyond the boundaries of Montreal Lake Cree Nation traditional lands. However, with the sponsorship of Weyerhaeuser and the forest management resources of Saskatchewan Environment, PAMF served as a classroom for a wider public body in which to foster the idea of sharing and developing values that rose above just logging and trapping of furs. More folk became involved in voicing their needs. The presence of Prince Albert National Park within the PAMF envelope expanded the opportunity to integrate a Park philosophy into the mix.

Weyerhaeuser had already initiated a stewardship committee wherein the Company's future operating plans were explained and feedback obtained. Where are the roads to be built? Where are the cutblock boundaries located? What impact do these have on viewscapes? What compensation is arranged for the trapper who's trapping block lies within a cutblock or two? How will Weyerhaeuser accommodate the needs of an adjacent fishing lodge that, until now, was only accessible by float plane? Etc., etc. Management foresters find these meetings heavy weather. It is wise to bring along a facilitator. Within the FCSask mandate all this becomes the responsibility of the Assistant Chief, Stakeholder Facilitation.

Once policy ingredients are agreed with, the process requires the development of a strategy to achieve the policy goals identified. Finally, an action plan is created, discussed, pulled apart, put together again, then implemented.

PAMF is more of a model that works within a regional context. And, its components are easily absorbed into an FCSask model. All we need now is a leader and political support together with a technical development team with vision.

You probably understand that the foregoing is set down to stimulate the development of an improved forest administration for Saskatchewan – much need for further deliberations before FCSask becomes a fact – and its need for a PAMF, Hudson Bay MF, Meadow Lake MF or La Ronge MF? Is one a duplication of the other?

And remember, all FCSask staff contribute to the bottom line.

B. Proposed Forestry Commission BC (FCBC)

In contrast to the Saskatchewan model I conceive of a devoluted forest administration in British Columbia. First, replace the Forest Service with an FCBC HQ in Surrey (or Kamloops). Second, deliver forest-based programming through consultations with regional districts and municipalities. Put forests back in the hands of the people that own them. Keep them up to date with forest practices, road use, and forest regeneration. Work with recreational groups to optimize the needs of various constituencies, including watershed management. Add road user fees for logging traffic on regional/municipal roads. Third, add a ministry with specific responsibility for drinking water. Gazette the watershed boundaries and exclude them from logging. The first thing this does is claw back all those watersheds purloined by the Forest Service in 1982 or so.

Second, involve regional districts and municipalities in forest planning. Contract out all forest-based services; ensure that each new management unit maintains its wood fibre supply obligations to the forest tenures inherited. Adjust administrative boundaries to comply with forest administrative boundaries – use watershed or heights-of-land boundaries. Blend this in stages with the existing tenure system over time. Within ten years, the intent is for forest operators to purchase their timber directly from the regional district or municipality. Each regional entity collects timber dues originating within its borders. It is a big job to set all this up. But think in terms of a century not a government four-year term.

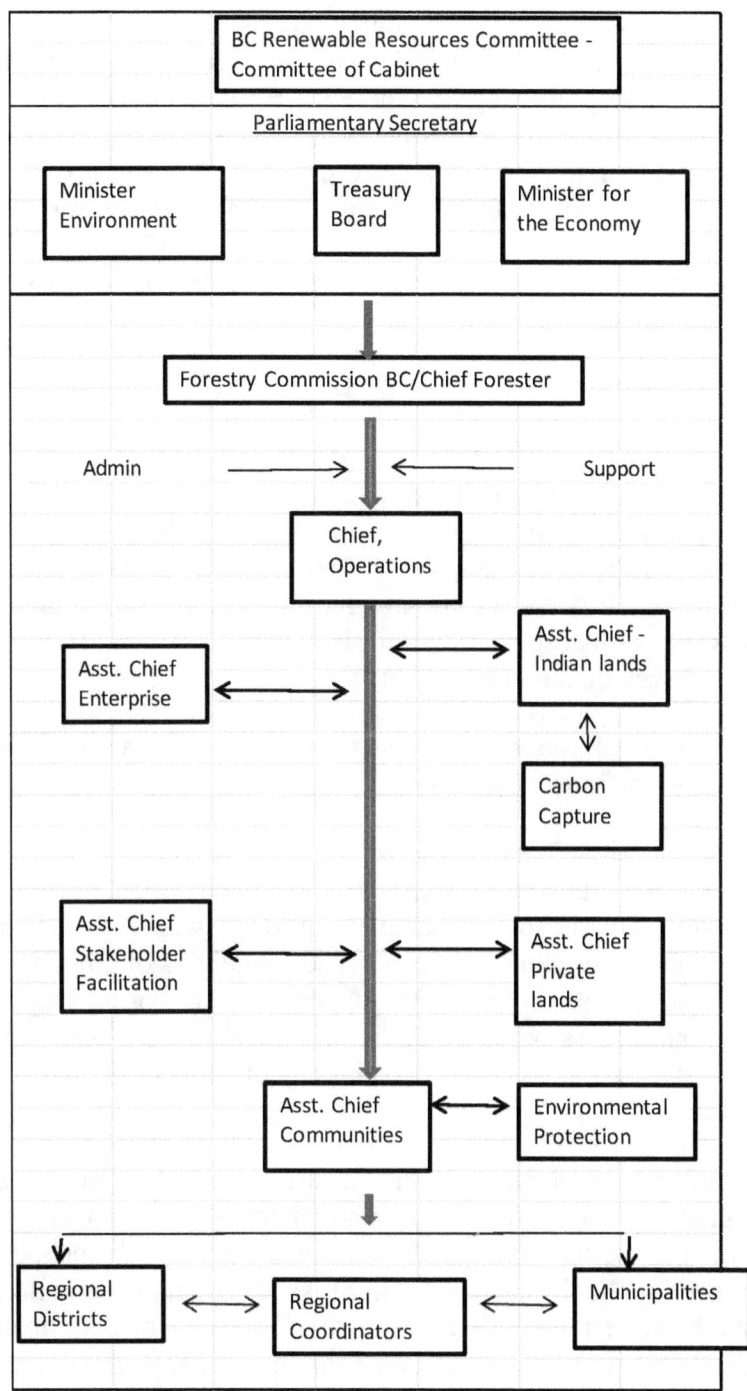

Figure 52: Organisational framework for Forestry Commission BC

The regional district or municipality will undertake all forest management duties using a core staff and contract professionals. FCBC will supply all support. The PAMF model works well at the regional level. Stakeholder involvement strengthens and provides for a more recognisable pathway connecting various 'interests'. As with Saskatchewan, accrued timber revenues and fees in the form of net dollars after expenses, will find their way to Victoria via the FCBC forest administration unit involved. All forestland will be restocked to FCBC standards.

Ω

Undoubtedly, at least one significant hurdle remains – the Association of BC Forest Professionals. Professional foresters and forest technologists work for government or industry, for the most part. Many of them will become the providers of forestry and allied services to FCBC directly through a fee-for-service system. Industrial foresters will probably continue in place. However, they will answer indirectly to regional municipal entities.

Again, the Saskatchewan Association of Forestry Professionals will undergo the same adjustments.

Another significant hurdle is the need to adjust administrative boundaries to remove uncertainty involved with overlapping jurisdictions. This is so noticeable in Saskatchewan with fur blocks. These should have boundaries controlled by heights of land, drainage systems and access. In BC, the rugged terrain, fjords, large river systems and access will create issues as municipal and regional district boundaries become a subset of forest administration boundaries.

Until a strong leader with vision surfaces (a 'plus tree' in human guise) matters stay the same. The whole purpose for this change is to grow the forest estate and transfer the forests back to the people. The government's 'Small Business Enterprise Program' was ill thought-out. We have a public asset. It requires a corporate approach to administration to achieve targeted outcomes. May my crucifixion commence?

Lastly, I realise that carbon capture in forest management will require research before it is adopted as a function of a Forestry Commission. I have been aware of its potential since my Enfor project days. I am also aware of the vast release of carbon into the air from forest fires in both Saskatchewan and BC.

Ω

Chapter 34:

Indians (*Candians*) and forestry

Do I stay with the label 'Indian' as advised by Chief Roy Bird in 1985? Or, do I promote a more inclusive label *'Candians'* without jeopardising First Nations' sovereignty.

I am ashamed of conditions on our Indian reserves. I am ashamed of conditions in the Vancouver downtown east side, I am ashamed of living conditions in parts of Regina, Saskatoon and Prince Albert. Crime breeds and flourishes in these ghettos. As Dale Reid observed, many young men and women from the reserve are in and out of jail regularly until the age of thirty. After that they settle down. Do you remember the Avenue Hotel on Central Avenue in Prince Albert? Right across from City Hall and kitty corner from my first office – murders, beatings and drunken brawls on a regular basis. A conscientious effort to involve *Candians* in achieving benefits from forest resource management on traditional lands mitigates the damage done to their culture and quality of life.

I have wrestled with this dilemma through much of this book. Why does it matter anyway? Because, within the public-at-large, the term 'Indian' lacks clarity and is too generic. For one thing, this label does not differentiate between an Indian with a treaty card (First Nations) and a non-status Indian or Métis, or an East Indian, or a West Indian. In BC, Indians are registered in lieu of a Treaty card. This lack of sensitivity by the public reduces political kinetic energy directed at improving Indian quality of life within the Indian envelope. That is why we have 'Idle No More' demonstrations as the Indian world attempts to market its needs and aspirations. We have stolen its birthright and lands. Isolated and abused, many Indians have yet to become part of the mainstream. This is not a static circumstance. Indian artists, lawyers, teachers, judges and those with other skills move through that often difficult societal barrier and coexist with the rest of society. I said earlier that it will take three generations or more to heal the damage done by the residential school system. It will take all of this to get rid of the societal damage contributed by the Indian Act as well.

Minorities are stomped on forever by our larger society. The Canadian government's solution is to build more and bigger prisons. How utterly short sighted and reactive. *"While Aboriginal people make up about 4% of the Canadian population, as of February 2013, 23.2% of the federal inmate population is Aboriginal (First Nation, Métis or Inuit). There are about 3,400 Aboriginal offenders in federal penitentiaries. 71% are First Nation (First Candian) – 24% Métis and 5% Inuit"*. (Wikipedia). The minute you exchange the term 'aboriginal' for *Candian*, the entire flavour of the discussion alters. Why, they could be my neighbours. What and who has put these folk in jail?

I abhor the 'aboriginal' label for First Peoples. It is an ignorant 'put down'. We were all aboriginals, and some still are. The label 'indigenous' is amorphous.

In the 70s my dad, in his chief medical officer's capacity within the BC Attorney General's Department, organised the startup of the Chilliwack forestry camp for young prison inmates (Oakalla). This concept spread into other jurisdictions. Give a youngster something to do then observe gratifying results.

En.wikipedia.org/Indigenous_peoples – *'Indigenous peoples are those groups especially protected in international or national legislation as having a set of specific rights based on their historical ties to a particular territory, and their cultural or historical distinctiveness from other populations."*

But we got it all wrong when we relied on the Indian Act of 1876 to legislate specific rights based on land use associated with Indigenous Peoples. Instead, we White people restricted them to reserves and imposed a patriarchal regime delivered through Health, Education and Housing. Indian self-government was replaced with a bureaucracy that became the lid to a large container. Until 1960 in

Saskatchewan you could not leave a reserve without permission from the Indian agent. I have just finished reading Alan Morley's biography of Peter Reginald Kelly. Peter was the grandson of a heriditary Haida chief. Peter's mission in life was to help raise the level of Indian quality of life to that of the surrounding Whites through 'association'.

We have to understand why the dilemma between them and us. The only solution the federal government endorsed in 1897 and earlier was to herd Indians into reserves and residential schools and encourage them to concentrate on agriculture. This was fine if you weren't Woodland Cree or Dene, or in northern Quebec or on Hudson Bay. They depended on the land for nourishment. This disappeared as the settlers encroached and the flourishing fur trade suffocated food. Starvation became a fact of life. And the settlers took over farming opportunities on reserve on a rental arrangement. Then the transfer of the Indigenous landbase from federal to provincial jurisdiction in 1930 sealed the fate of Indian Nations throughout Saskatchewan and Alberta. In other jurisdictions this occurred in other ways.

The problem is that Indian homelands relied on the Fur Trade and buffalo. These evaporated one hundred years ago, leaving behind a *Candian* residue. Homeland communities are rural, remote and without cashflow to support a reasonable quality of life. They lack the job skills, learning and mentoring. In steps the Department of Indian Affairs with its patriarchal mentality. The Indian Act treats them as wards of the State, rather than a part of Canadian society. This has throttled any attempt to develop a stronger association with Canada-at-large.

First Candians are wary of outside intrusions to their lifestyle and culture (sovereignty). They lost so much independence over the last 150 years. This includes the erosion of their place as 'First Peoples'. They defy any attempt to integrate into the larger society. Any grand chief of the Assembly of First Nations loses his job after cozying up with any part of Canada's political empire.

Ω

First Nations evolvement had three optional paths:

451

ONE Leave the shelter of an Indian roof, take the white man's scrip and become non-status.

TWO Stay on the reserve, continue to accept the handouts from Indian Affairs, but live in poverty.

THREE consciously educate yourself, learn skills and become a citizen of both worlds.

My *'a priori'* estimate from my work observations with Cree and Dene would suggest a 30:52:18 ratio. First Nations leadership exists in the THREE PLUS world. It benefits from language, education, business and networks. This will end with 'association' with Whites rather than 'integration'. No First Nations chief will vote for an AFN[122] leader promoting integration.

It is time to throw out the Indian Act, and consider the following:

My first concern, after working so closely with various communities on forestry and other matters, is that we treat the Indian side of our Canadian population with so little respect and inclusivity. Let us all reject the Indian label and now refer to our Indian confreres as *'Candians.'* We have Brits, Ukrainians, Germans, Chinese, Poles, South Asians, East Asians, West Indians, East Indians, and Japanese – all Canadians. So why not include *Candians*. This denotation would take down the fence to some extent. Wikipedia records that 4.3% of the Canadian population is Canadian Indian, but they occupy 36.5% of the landbase. I go one step further: in recognition of the fact that some *Candians* are 'First Nations' and others are 'Northerners' or Inuit, we need subsets to identify First Peoples more particularly. I modify this label.

Technically, *'First Candians'* is limited to those that have treaty cards or registration, and otherwise legal status on the landbase. Residents on their reserve have control over the day-to-day functions inherent in living within a homeland community. Exit the label 'reserve'. *First Candians* will manage traditional land parcels in concert with other interests in the future. Within the sub-treaties this happen through partnerships with either government or commercial

452

enterprises, or both. On the reserve, land is free of valuation. It is owned by the homeland community and is not for sale – keep it that way. One re-emphasises the Natural Resources Transfer Act of 1930. Most of the Saskatchewan and Alberta land base was transferred from Canada to Alberta and Saskatchewan. This destroyed the rights of Indian Nations to the use of their traditional lands for food. Coincidentally, this has damaged Indian society as much as the residential schools debacle.

An example of this is the failure to integrate provincial lands into business partnerships. Both Perter Ballantyne Cree Nation and the seven nations cooperating to form the Island Forest resource lands launched a partnership with the province northeast of Prince Albert. Work started on this in 2002. According to the government website, the Ministry of Environment continues to posture and delay the completion of the process.

BC has a more complex situation brought about by a historical refutation of any special status for Indians. In the early days the informal policy was to eradicate them. Then came the benign years when Whites stole timber, fish and food. Recent governments have attempted to assuage their guilt with good shows of negotiation. When these look like achieving success, the province immediately cancels any further efforts. The only two successful outcomes came from the Supreme Court of Canada. Unfortunately, this seems to say a lot about British Columbians. Disgraceful really.

BC forestry professionals need to recognise *Candian* goals of 'association' and not 'assimilation'.

In conclusion, the Indian Act creates a split between 'the haves' and 'the have-nots'. This creates a dichotomy between those that gain from the Indian Act and those that leave. Reserve Indians are loathe to leave because housing is subsidised and Indian Affairs pays a higher welfare rate than the province. Also some chiefs want to hang on to the Indian Act because they are able to tap into the cashflow available.

Candian sub-treaties of tomorrow

There is a 'duty to consult' with *First Candians*. When encroachment on to the traditional landbase is contemplated the instigator has a duty to perform. The recent Supreme Court decision has reinforced this ethic. Up to now, the process has appeared to be more of a lip service than a sincere desire to develop parallel interests. This is gradually changing. However, each encroachment should be mutually covered within a sub-treaty to the original main event. There are many unscrupulous interests still roaming around, including governments. Negotiate a sub-treaty – ensure that real benefits accrue to the *Candian* communities impacted. The sub-treaty should include clauses to maintain environmental stability on the traditional land base affected.

Any mine, forest operation, pipeline and any other land use impacting *Candian* traditional lands will proceed through a sub-treaty process. In fact, *First Candian* treaties and sub-treaties require formal associated rights within a universe parallel to 'The Canadian Charter of Rights and Freedoms'. *Candians* have the right to say 'no'.

The framework for a sub-treaty

Sub-treaties will replace the Indian Act. They will be a subset of existing treaties. They will be negotiated between chiefs and councils, and the relevant province. Canada will continue to supply funding. The $5 per year honorarium per head now paid by Canada will be scrapped. The replacement will be an annual fee paid based on the hectares of traditional land registered to each First Nation. Métis and non-status communities will share in the landbase. Canada will transfer the money to the province. This government, in turn, will honour the sub-treaties and transfer the allocated funds to the first Nations and Metis communities. So much per hectare. The value per ha will be determined through cost-benefit analysis. But it has to be sufficient to cover the programs identified in the sub-treaty.

This fee also covers Indian lands now subsumed by White populations. Thus, communities, mines, farms, ranches, boat docks, other infrastructure and recreational may be included.

This land fee will finance the operation of homeland communities. These will adopt the provincial municipal acts and regulations. The annual fee will also pay into provincial Medical Services plans and education, including post-secondary studies and housing.

Where a dispute entails as to the amount of the fee, arbitration will settle the matter.

Contracts based on sub-treaties will dictate resource use and environmental mitigation. Any participating homeland community will be reimbursed through 'rents' for the use of the resources utilised. Monies from these sources will be paid into the homeland 'accounts receivable'. They are separate from monies passed along to and by government.

A few conditions

Homeland community members can participate in any enterprise contracts as staff, employees and subcontractors. Where insufficient numbers of workers are available in one community, *Candians* from other communities are considered for hire. Anyone may be considered for specialist assignments.

No chief or council member may receive monies from the enterprise perceived to be bribes, freebies, finders' fees and in-kind contributions. Legitimate personal expenses incurred on behalf of the community will be reimbursed. Each sub-treaty will cover all homeland communities identified within the original treaty, plus adhesions, plus new reserves created under the treaty land entitlement outcomes. Chief or councilors with involvement in the enterprise may have to resign from office.

Métis and non-status people already living in a homeland community will be absorbed into that community. Resident non-*Candians* will pay normal provincial municipal taxes and may vote in municipal elections.

There is a lot more to define, organise and invent of course, but the framework should float. BC should bite the bullet and use the same template for implementing *First Candian* sub-treaties. Running to the courts and waiting seventeen years for a decision is just too much. The first article in a BC-generated sub-treaty will be the definition of which traditional lands are to be included. That one task should work to a deadline, or we spend another 100 years waiting in suspense.

Here on the Sunshine Coast, Sechelt Nation has the first municipal governance in place anywhere on *Candian* lands. It interacts with the Sunshine Coast Regional District to manage certain lands it considers to have responsibility for. Land uses such as shoreland developments, boat docks and wharves tributary to the Salish Sea are part of its management responsibilities. Sechelt Nation also manages its historical sites.

Carbon capture/cap-and-trade

Trees capture and store carbon, as does undergrowth and humus layers, and riparian ecosystems. It is my desire to see cap-and-trade features set up on all *First Candian* and Métis traditional lands. The modelling is available to measure carbon sequestration over large areas using LIDAR and other tools. This has to be done one landscape at a time. Some carbon deficits caused by old age or fire may cancel a positive carbon sequestration value on another landscape.

Alberta farmers in a cap-and-trade relationship with oil companies get paid $15/ha annually for range lands capturing carbon. It is high time that *First Candians* put their traditional lands to work earning a return on the management of carbon capture. Many *Candians* will require training in forestry skills, data collection and carbon modelling. Where some forests are old and give up some carbon, other younger forests sequester carbon. In any event, it gives young people in those homeland communities a career path.

The science behind carbon capture is incomplete. Forest ecosystems will require sufficient data to run growth-and-yield models with acceptable precision.

A Retrospective

Candians are a proud people with skills, philosophies, survival acuity, and a belief that incorporates both environmental and spiritual values. They are our future, but we yield their place to work permit folk from overseas. I can just hear a fast food franchisee telling me that he has tried to hire 'Indians' but they do not show up. Well then, try harder. You will have to start with some life skills awareness and make sure that they have a place to stay, a mentor and transportation. I believe each new Canadian should cosign an existing treaty, or cosign a statement which recognises that *Candians* are our First Peoples and have rights that were given to them at birth.

After returning from Irkusk in southern Siberia, and noting the Mongol residue within the native population, I was fascinated to discover Mongol physiognomy within some Montreal Lake Cree Nation summer students. And, one of my long-term Hungarian forester friends, who was part of the great escape from Hungary in 1956, also has the Mongol eyes. We are all brothers under the skin. Bloody Canucks.

On one occasion, Montreal Lake invited me along to visit Carrier Lumber Co. in Prince George and a brand new wood pellet plant – a potential pathway into developing such facilities in Prince Albert. Carrier might partner with them, serve as a mentor, and otherwise create a pathway into the establishment of jobs, cashflow and other benefits close to the reserve. But, it was not meant to be. Chief never showed. He had discovered a casino. The rest of them were bored out of their skulls. The ladies went shopping.

Carrier and Tolko are two forest industry companies that take much trouble to incorporate *Candian* input into the workforce. Weyerhaeuser just paid lip service, though they obtained their raw logs from *First Candian* traditional lands.

Weyerhaeuser went out of it way to avoid any *Candian* involvement – most of the time. One exception was Wapawekka Forest Products Ltd in Prince Albert – a joint venture with the three Woodland Cree bands. But it was a sham really. Weyerhaeuser used it as a wood chip supply for the pulp mill next door. Indian Affairs and others paid for the capital costs associated with this venture. No one asked me for my opinion, even Gene Kimbley, my colleague and friend. He sat on the board as a representative of Montreal Lake. Weyerhaeuser Timberlands in its bifurcal way also sponsored summer student jobs. On one occasion, Larry Bird from Montreal Lake was hired for the summer. The Weyerhaeuser PR folks put his photo on the front cover of a Company newsletter. His IQ is outstanding.

I did write speeches for chiefs and attended national forest management meetings in both Vancouver and Ottawa. However, it was all so disconnected. More of a make-work project for the bureaucracy than anything else. We have to protect our pensions, and so forth. However, when I scan the National Aboriginal Forestry Association website, it seems to have come a long way since those early days. This gives me hope.

More thoughts

If I leave the reserve, I have to rely on provincial welfare, but it pays less than Indian Affairs welfare, so I stay on reserve. As a Métis at Pinehouse, Saskatchewan, Sask Housing only charges me a nominal rent for my home. But, if I get a job that creates payroll, the rent jumps. It does not pay me to take that job. Now I think of it, this is probably why Pinehouse walked away from its partnership within Weyotun Resource Management Ltd. If the boys make money, it is scooped back by increased rent. God help me – I had not considered that until this moment.

First Candians from Peter Ballantyne country and from Montreal Lake were the backbone of my field forestry crews throughout my Prince Albert years. They learned fast, their data collection and recording was top notch, they worked until 9 at night if this meant not having to return to the same site tomorrow; they were resourceful and could be comfortable in the worst weather. That reminds me of that February camp north of Melfort while on the poplar chopstick project. No sleeping bags, just blankets at minus 26 Celsius. Supremely uncomfortable. The one and only negative in all my experience hiring *First Candians* from the reserve was that they had no wheels. So, they could not get to work unless I picked them up. For Montreal Lake, that meant a two hour return trip from my office. Conversely, Delilah Bird, our GIS trainee from Montreal Lake, had to make the two- hour return trip to get back home at night.

I would drive 400 km; get to Pelican Narrows by 8.30 am for a meeting. The ladies in the band office always giggled from behind the counter on my visits to confer with Chief and councilors. They briefed me on the current 'happenings'. The council would finally call me in on the tail end of the day's agenda at 6 pm. There was that need to keep me in my place. They made up for it though by taking me with them on fishing trips and big tours of sawmills in British Columbia and whatever other excuse they could make up. I made up the itineraries. Dale Reid, Chief Ronnie's assistant, organized the logistics. The destination was always Vancouver.

Every community had a new band office, nursing stations did get replaced, sewer and water infrastructure did get upgraded, but the new schools were built too small. New hockey rinks were virtually destroyed before the second winter – ice-making equipment destroyed in the process.

But anyone with a job in the community had decent wheels, the same problems bringing up kids as elsewhere, and homes to keep up. By jobs, I include trapping, fishing, guiding, running the gas station and convenience store, and operating the community grader, gravel truck and garbage disposal.

However, another layer in the community lived in poverty: ten people per household, floors, walls and windows damaged, leaky roofs, mold, and graveyards of discarded possessions. It is difficult to remember that, until 1973, Pelican Narrows had no access road. Everyone travelled by boat or canoe to the nearest transportation hub. The people lived apart in their own domains, coming together once a year to pick berries, marry, trade information, pick up more supplies and get ready for winter.

I loved going to Pelican Narrows when the band office was still located there. Being winter, I was allowed to live in a cottage at Mistanosayew while carrying out various forestry training projects. Eventually, I was advised against going to Pelican – too much potential hostility. But I did stay in Sandy Bay in later years at a new senior's domicile. It was still empty and consisted of a row of brand new cottages.

I have yet to mention drink, drugs and abuse and fighting. There is also history on those reserves going back generations. The Halters hate the Balladones; Crusters have been at war with Michaels forever. It means that someone gets a new house and someone else doesn't, someone gets a job driving the garbage truck and someone else doesn't. In the old days, a group of dissatisfied families left the community and set up elsewhere. The creation of Deschambault occurred like this. Today, though, all have to stay put, except that now they move to the city. Do you remember the situation at Montreal Lake where we had to count annual tree rings as part of a research project? In some years these rings were so narrow that they were almost impossible to read. I informed the crew that those were the bad years associated with a bad chief. Poplar trees cannot leave. They have to stay put. Or, Indian Affairs scoops the band's housing budget to make up shortfalls in other areas.

All I am doing right now is skirting around the solutions. I have none. But, let's throw away the name 'reserve'. It has so many negative connotations. Instead, let's refer to a reserve as a *First Candian* homeland community such as Pelican Narrows. The village of Beauval, SK, is a Métis, or *Candian* community. Like elsewhere, these are good, safe places to bring up kids. The parents have to have income, the grandparents take a large role in rearing the youngsters, and teenagers need mentoring, supervision and something to do. In earlier times they went to the trapline, caught fish, and were taught to hunt.

This is all why I tried to introduce integrated resource management in these communities. Remember every 525 m3 of logs (10 truckloads) supports one full time job in small scale forestry. When we include ancillary jobs in all aspects of forest planning, preparation of cutting plans, environmental monitoring, forest fire protection, tree planting, fencepost manufacture and so forth, young *Candians* will stay out of gaol, and a healthier community is born. The rule of thumb in Pelican Narrows was that a youth was in and out of gaol between the ages of 18 and 30. After that, he, or she, settled down. Gestation may take years, but I guarantee anyone that we will notice results within six months of startup. However, this all has to be the responsibility of the Indian world. I have no business even voicing these thoughts.

What do we need to initiate this co-op forestry initiative, community based?

- All revenues from marketing and sales of products manufactured will remain with the co-op.

- Employees to be prepared to leave the community to receive training in specialized skills.
- Employees should graduate in life skills prior to going to work for the co-op.
- Chief and council not needed in the operation of the co-op. It operates at arm's length from band (community) administration.
- Women have equal opportunity as do men. Women are better managers anyway.
- Co-op to provide scholarships to those that seek higher learning.
- Jobs include fisheries officers, tourist and recreation staff. GIS mapping, ornithological studies and inventory, carbon capture reports, band history projects for sharing with recreationists. The Sturgeon – Weir River is steeped in indigenous history and the travels of early explorers. It was the link between Montreal and Rupert's Land, or through Frog Portage on to the Churchill River and Hudson Bay. "Mofries, you can do it, but stay focused, eh?"
- Co-ops need mentors akin to the CESO program. (Canadian Executive Service Overseas).[123]
- Arm's length involvement from Indian Affairs. No Band Council Resolutions . Eventually, Indian Affairs should disappear.
- Co-ops need empathetic integrated resource management agreements with the provinces.
  Candians need to manage their traditional lands. In British Columbia, the first duty for all *First Candians* is to mutually agree to tribal boundaries.

In conclusion

I watch Grand chiefs from the Assembly of First Nations turfed on a regular basis. These are leaders with impeccable credentials. To me, they beat any prime minister we have in office. Somehow, we need to assign seats in Parliament to elected *Candian* representatives. They will represent *Candians* and vote on legislation. There should be at least three per province, but based on population.

461

What can I say? *Candians* have sovereignty over their lands. I had no business penning this chapter.

Postscript: a primer for forestry professionals in an Indian world

Assumption: Indian timber close to the mill becomes really attractive to the company forester. The timber is adequate in terms of quality and volume. It is also low cost as delivered to the mill. The truck haul passes through Indian lands.

1. Contact chief and council. Ask for permission to attend a meeting to explain your needs. After the meeting make sure you have the name of a key contact on council as a follow-up connection. Do not become impatient. Ensure that you are high enough up in your hierarchy to negotiate on the Company's behalf.

2. Follow up by asking the key contact and/or the chief's executive assistant for a list of young folk from the band who would like to become involved in a forest operation on their land. Do they have wheels? Could they camp to be close to the work? My recommendation is not to proceed too far until you have a chance to discuss your circumstances with foresters who have some familiarity with your mission. If you are stuck, phone the woodlands staff at the Tolko mill at the Pas, MB. I was so appreciative of the assistance provided to Gertie Budd and her community at Sturgeon Landing, SK on the task of rescuing prime white spruce blowdown on their reserve.

3. Get your HR folks to apply for training money, sustenance and a training wage. Your forestry department will need to assign a trainer. Make sure there is money for safety gear and cruising vests. Perhaps there is a local person with Indian ancestry available.

4. Involve your new crew, both men and women with everything from stump to dump. Start with the timber

cruising and mapping. Bring them into the office to crunch data and assist with planning the layout.

5. Get them to flag the cut boundaries lay in the roads and landings.

6. Perhaps one or two could train as machine operators.

7. Tree planting should become part of further training. Ensure that there is money for dibbles, bags and tarps – plus transport, plus arranging for the supply of seedlings. Issue the band a contract for planting.

8. Now the crunch. Your company pays stumpage to government, but also prepares to negotiate a rental fee per cubic metre to the band. At Sturgeon Landing SK Tolko paid $5/m3 to the band. Also, ensure that the money goes to the band and not to Indian Affairs in trust. This latter agency will scoop it for general revenue.

9. The payoff is long-term. Train Indians to do the PHSPs, waste surveys and regeneration surveys.

With the foregoing, I come from a Saskatchewan experience. I remain totally perplexed by the BC Indian scene. However, I am mindful of recent Supreme Court rulings with respect to Indian lands in BC and the needs of resource industries.

ΩΩ

A request

For those who have read this far:

Send me an issue or message that you believe to be priority 1 in this book.

I may be offended, but will live with the consequences. Please limit your comments to the content.

We may even launch a second edition of 'Silviba' and incorporate the remarks tendered.

I have built a website www.Silviba.com.

43-1413 Sunshine Coast Hwy

Gibsons, BC

V0N 1V5

Canada

Silviba@eastlink.ca

ΩΩΩ

A map displaying timber berths and Indian reserves tributary to Prince Albert in 1900 will be filed with the Prince Albert Museum. Hopefully, a second copy will end up with the Saskatchewan Archives Board in Saskatoon.

466

Index

Silviba endnotes

Chapter 2
[1] The air photos arrived with boundaries drawn around major forest species ecotypes.
[2] A closed boundary containing trees with the same characteristics.
[3] Determination of the annual allowable cut (AAC) on a Sustained Basis for Virgin American Forests. E.J. Hanzlik, U.S. Forest Service. Journal of Forestry, 1922. Google Hanzlik/forestry
[4] BC Timber Sales is a provincial government agency charged with raising $60 million annually from Crown dues.
[5] Diameter at breast height, 1.3 metres above average ground level – a standard measure for tree diameter outside bark
[6] Chain = 66 feet; 80 chains = 1 mile An abney measured the percentage reading to the top of the tree plus the reading to the base. Add the two readings together, then at 2 chains distance from the tree, multiply the total by two = total height of that sample tree. Today tree heights are taken with digital height measuring equipment.
[7] *Oplopanax horridus.* The curse of all timber surveyors. Poisonous thorns all over the plant. They love knees and shins.

Chapter 5
[8] Peter Bentley has written a biography of the families. He has far more to tell than me.

Chapter 6
[9] A light wire rope cable used to pull heavier cables of a rigging system used to skid logs.
[10] Glen ended up managing all Canfor operations in Alberta. He was Roy Jewesson's predecessor in the Nimpkish Valley.
[11] Relative humidity, an indicator of moisture content in the air. The critical value is 30%. Below this, sparks cause fires.
[12] Set of fuel moisture sticks – Douglas fir dowels 1/2 inch diameter were placed on a wire support just above ground level. The hooktender measured the weight of these periodically with a big brass backed scale. Weight was calibrated as a percent of the stick weight taken up by ambient moisture. Once they fell to 10 %, the logging operation was to cease. Hands had to be free of grease and clean while handling the sticks.

Chapter 7
[13] Tree seedlings were planted as 2-year olds with no intervening transplanting.
[14] Trees with similarly observable characteristics, shared genotypically.
[15] K line is a reference to the Nimpkish railway, otherwise known in Indian as *Klaanch.*
[16] H line or Hoomak logging railway was used to access the Davie River valley.

Chapter 9
[17] Jack was my counterpart at Woss camp.
[18] Bentonite, a brightly red retardant dropped from water bombers, had a clay base. It coated forest fuels and retarded combustion.

Chapter 10

[19] A gated fabric clad container suspended below the helicopter. The bucket could hold 200 gallons. The pilot scooped water from a lake or pond to refill the bucket each time. The pilot controlled a switch that opened the gate and allowed water or retardant to bomb the fire. Today this system has become more advanced.

Chapter 11
[20] My term
[21] Each forest jurisdiction or forest tenure attempts to maintain a balance between timber depletion from logging, insect damage or diseases, and the rate of regrowth. One element of maintaining this balance is to replant logged area and maintain the allowable cut.

[22] Google.ca (cfs.nrcan.gc.ca/pubwarehouse/pdfs/10068.pdf)

[23] NTS = National Topographic Series (Natural Resources Canada)

[24] NSERC (National Science and Research Centre Canada)

[25] The *SKY GENIE*® allows a preset rate of descent by number of turns around the shaft. Few turns for rapid descent, more for slower rates.

[26] Rope anchors used in a number of situations requiring the tether of a person to a rope.

[27] IFFS crews referred to themselves as 'sky spiders'.

[28] Columbia Cellulose held Tree Farm licence 23. It covered the area from Castlegar in the south, to its northern limits just south of Kinbasket Lake – the entire arrow lakes drainage.

[29] Not more than 2 metres above the forest canopy. The lower the hover, the faster the crew got to the ground and released themselves from the rope. As long as the crew was attached to the rope, the pilot was severely constricted as to his movements.

[30] DAP can be used as a fire retardant. It lowers the combustion temperature of the material, decreases maximum weight loss rates, and causes an increase in the production of residue or char. These are important effects in fighting wildfires as lowering the pyrolysis temperature and increasing the amount of char formed reduces that amount of available fuel and can lead to the formation of a firebreak. It is the largest component of some popular commercial firefighting products. (Wikipedia)[

[31] The area is the upper end of the Lardeau district. The river's course is nearly due south from its origin in the Selkirk Mountains at the southwest toe of the Illecillewaet Neve, which is on the south side of the Rogers Pass and is the source of the Illecillewaet River. Sometimes called the Fish River, this is a wild outfall, amid large cedars, hemlock, devil's club and bears (Wikipedia).

[32] A plywood step attached to the skid to reduce ground pressure in soft spots.

[33] VFR or Visual Flight Rules

Chapter 13
[34] This council was not formalized until 1985. However, in 1982, the mold was largely in place.

Chapter 16

[35] The Crown share fluctuates according to market pricing.

Chapter 17
[36] This number changes throughout the narrative. I cannot find this directory.
[37] The Natural Resources Transfer Act of 1930 transferred all resource lands within the former Northwest Territories to provincial jurisdiction. This removed any implied rights to land use by Indians, except within gazetted reserves. Saskatchewan never was a party to the treaties.
[38] http://www.peterballantyne.ca/
[39] http://www.peterballantyne.ca/
[40] ww.otc.ca/siteimages/treatymap_large.pdf
[41] Economic development
[42] These funds originated in PBCN "head office" via INAC – Indian and Northern Affairs Canada.

[43] March 13, 1989
[44] Refers to screefing a cutblock prior to tree planting
[45] Alan Appleby killed this initiative. He intended to turn over Indian lands to an outsider.

Chapter 19
[46] A backpack can holding water, together with an attached hose and hand- operated pump. It holds about 5 gallons.
[47] Not entirely true. Large swaths of conifer did regenerate in areas around what is now Prince Albert National Park, formerly the Sturgeon Forest Reserve, established in 1912. The mixed poplar – white spruce ecotype originates from this fire history.
[48] Ladder Lake Lumber Co. in Big River (Bodmin) was burned to the ground in the 1919 fires.

Chapter 20
[49] Preharvest Silviculture Prescription sample plots.
[50] Geographic Information System used to display area visually based information, amongst other attributes.
[51] I mentored both Dale (Peter Ballantyne Cree Nation) and Gene Kimbley (Montreal Lake Cree Nation). It was a matter of 'training the trainers', who then advised chiefs and councils on forestry opportunities on reserve as identified from time to time. A few councilors were antagonistic to Silviba and were suspicious that the 'white' people were stealing from them.
[52] ENFOR = Energy from the forest.norcan.*gc.ca/publications? Id=19379*

[53] A stratum subdivides a population, human or inanimate, into clusters of similar features. This reduces the need for sampling since each cluster is homogeneous.
[54] Black spruce is shallow-rooted. In a subhygric regime, the soil is clay. Tree roots float in this impervious layer and a windstorm can roll back a forest of trees just as you would roll up a carpet.
[55] I was just guessing, but the minimum measures feel right.
[56] The prime mover was NorSask Biomass Ltd based in Meadow Lake. It formed a partnership with Beaver River Community futures to build the cogen facility.
[57] The price 'factory on board'. Today this means the price at the mill. The purchaser pays the trucking.

481

Chapter 22
58 Wind direction is always 'from'.

Chapter 23
59 The hewsaw was designed in Finland to process small wood into lumber. It incorporated a high-speed log infeed and suited Saskatchewan's boreal forest timber characteristics admirably.
60 Saskatchewan Environment and Resource Management in Meadow Lake. Floyd Wilson, the district forester had connived with Mistik to allow them to log on the ProForest timber permit.
61 Finger jointing consists of converting short pieces of lumber into two by lumber through a process of gluing these pieces together to form dimension lumber. The joins are shaped like fingers on two hands that then push into each other to form a bond. This has reduced sawmill waste significantly. ProForest estimated that it could deliver lumber shorts into Kalispell economically.
62 Peter Ballantyne's economic development director located on the urban reserve in Prince Albert.
63 Alphonse was a Montreal Cree Nation councilor, who went on to be chief, before ending up as Grand chief for the Federation of Saskatchewan Indian Nations. In due course, Alphonse, in all his sartorial best (a Grand Chief's headdress included) sat beside the queen at the opening of the Indian university in Regina
64 http://www.esri.com/software/arcgis
65 Saskatchewan Government Growth Fund

Chapter 24
66 Both the direction of swing and the opening and closing of the tongs were hydraulically controlled. During super cold nights, the operator could blow an O-ring and have to shut down.
67 See Western Boreal Growth and Yield (WESBOGY) Association ...
www.rr.ualberta.ca/.../WESBOGY/.../Wesbogy/.../2007DataCollectionMa...
Sep 3, 2007 - Western Boreal Growth and Yield (WESBOGY). Association. Long Term Study (LTS) of growth and development of mixed stands of spruce and.
68 Diameter breast height
69 Studs are small diameter trees bucked into 16-foot lengths. They convert in the mill to 2X4's etc.
70 Esri = Environmental Systems Research Institute. Developed GIS = Geographic Information Systems. Used to take spatial data and plot it on maps. Now the ultimate forestry tool.
71 Esri Canada www.esri.ca/

Chapter 27
72 A German consulting agency, my direct employer.
73 Aerial Forest Protection Service (Russia). Wikipedia, the ...
en.wikipedia.org/wiki/Aerial_Forest_Protection_Service_(Russia)
The Aerial Forest Protection Service (Авиалесоохрана, or Avialesookhrana) is a Russian government agency charged primarily with the aerial management of ...

[74] National Oceanic and Atmospheric Agency (based in the US). This agency operated a global weather satellite service.

[75] EU/Technical Assistance to the Commonwealth of Independent States

[76] Canadian Interagency Forest Fire Centre, based in Winnipeg.

Chapter 28

[77] Saturday Special from the Vaults: A History of Cumberland ...

edwardwillett.com/.../Saturday -- special -- from -- the -- vaults -- a -- history -- of -- cumb...

By Edward Willett -- Here is something I wrote for Cumberland House a few years ago. ... Small groups of Indians began arriving at Cumberland House to trade moose meat, fish and ...

[78] Hunting Technical Surveys Ltd, the British general contractor for this Russian forest fire management system improvement study.

[79] Aerial orthophotogrammetry or digital rather than film imagery is used to update GIS mapping. It is both time saving and more accurate. The old ways incorporating laborious ground truthing were only used in special circumstances. Remember, the working platform for updating mapping was a digital image. At times, one had to travel to a specific site to match the imagery with the correct interpretation of the ground features *in situ*.

[80] WESBOGY: Western Boreal Growth and Yield (WESBOGY) Association ...

www.rr.ualberta.ca/.../WESBOGY/.../Wesbogy/.../2007DataCollectionMa...

Sep 3, 2007 - Western Boreal Growth and Yield (WESBOGY). Association. Long Term Study (LTS) of growth and development of mixed stands of spruce and .

81 WESBOGY: Western Boreal Growth and Yield Cooperative.Google.ca.

[82] Dbh is a standard measure of the diameter of a tree outside bark at 1.3 m above average ground level, in cm.

[83]Forest Ecosystem Classification project, www.environment.gov.**sk.ca**

Chapter 29

[84] http://www.google.ca/

[85] It is frowned upon to export roundwood rather than lumber, since not only does wood fibre leave the province, but so do the potential manufacturing jobs. That goes with it. British Columbia is particularly bad for this, since the forest authorities there allow green timber to be exported to other countries.

[86] www.environment.gov.sk.ca/.../adxgetmedia.aspx?

[87] Saskatchewan Institute of applied Science and Technologies, Prince Albert campus.

Chapter 30

[88] It is said that Henry Kelsey could have taken a side trip and reached the future site of Prince Albert in 1692 on his second voyage. He was supposed to have gone south down the Kelsey trail to Minnesota via the Carrot River.

[89] James Nisbet the man; but the Nisbet forest. = eternal confusion for those who wonder.

[90] http://www.pamodelforest.sk.ca/pubs/PRINCE ALBERT MODEL FOREST3200.pdf

[91] Website/Montreal Lake Cree Nation.

[92] From research conducted by Jason Nelson for ECGI at Little Red reserve in 2004.

[93] Saskatchewan archives/ Prince Albert district correspondence, January 1903.

[94] 1911 dollars

[95] Timber berths were the end result of a timber leasing arrangement, renewable after four years together with annual rent, and providing quarterly returns for all lumber and other products produced. After all the timber was logged, the berth reverted to the Crown.

[96] An experiential number derived from a timber berth volume analysis. It ranged from 2 to 7 million board feet per square mile as reported in the NR5 records. This was net volume in Prince Albert after losses from the river drives. Up to 25% for some timber berths, according to records.

[97] Saskatchewan archives, Saskatoon.

[98] Just read both the Department of Indian Affairs archives in Ottawa and NR5 covering the Sturgeon Lake timber surrender that was settled and the little Red timber surrender that remains unsettled.

[99] Just follow the money.

[100] Reference = Sask Archives NR5/just follow the money.

[101] You can read his letters to various authorities in the Indian Department archives.

[102] At that time known as Halcro community.

[103] I cannot find the document summarizing Chaskastapaysin's situation as presented in evidence at the Indian special claims enquiry brought to the table by James Smith Cree Nation. It was probably removed from the website.

[104] A land surrender claim brought to the Indian special Claims Commission remains unresolved.

[105] library This information, copied from Sessional Papers, Canada Department of Interior, are housed at the University of Saskatchewan

[106] Source: Indian Affairs timber officer, ex Indian Affairs archives, Ottawa.

[107] U of Saskatchewan archives

Chapter 31

[108] Martha McCartney, Harvesting the Northern Forest, The Pas History & Heritage Society Inc., Sam Waller Museum, The Pas, MB. (ISBN 1055056 -- 524 -- 9).

[109] From report of the Crown timber agent, Dominion Lands office, Prince Albert, July 6, 1906. Sessional Papers. Department of Interior, 1907.

[110] Lath is a secondary product in the sawmill and was used to hold the plaster on ceilings and walls before the days of gyproc. This product was not considered in the Sturgeon Lake timber surrender, but does become a significant feature of the Little Red timber surrender claim. It required an annual lumber output of 60 million board feet to produce 12 million feet of lath. Lath is a lumber byproduct.

[111] From report of the Crown timber agent, Dominion Lands office, Prince Albert, July 6, 1906. For convenience purposes, this wood supply is referred to as the 'Sturgeon Lake wood Basket'. Not only did this include the timber from the timber berths located within its boundaries, but also the potential timber volume from two Indian reserves, IR 101 and IR 106A.

[112] Until the arrival of PAL the three main licensees in the Sturgeon Lake wood basket were James Sanderson, sawmill owner, 10 berths totaling 87.14 square miles; William

Cowan, lumberman, 7 berths, totaling 59.26 square miles; George Burns, banker, 11 berths totaling 249.75 square miles. In aggregate, 396.15 square miles supporting an estimated sawlog inventory of some 1,981 million board feet of sawlog spruce inventory. PAL had assumed the administration of all these berths within a year.

[113] Remnants of the dam are still visible today.

[114] Estimated photogrammetrically from James' photographs, 1907 winter, in the Saskatchewan archives, Regina.

[115] Today's forest cutting boundaries show up on digital aerial photos (orthophotos). Trespass cuts are easy to detect on these photos using GIS overlays.

[116] This volume is more or less, what the investigation team estimated from IR 101 on the north side of Sturgeon Lake.

Chapter 32

[117] Department of Regional Economic Expansion – a Saskatchewan /Canada economic development agreement, expired in 1981.

[118] Manufactured by Prince Albert Foundry.

[119] The outside slabs sawn prior to making cants. These in turn became dimension lumber. .

[120] Slabs and edgings are resawn to create boards specialty lumber sizes, often yielding a better grade of lumber.

[121] FCS website

[122] Assembly of First Nations

[123] See CESO.

Questions for discussion

Some suggested responses will be posted on the Silviba website.

1) On a scale of 1(exceptional value) through 5 (a total time waster), was reading Silviba a good use of your time?

2) If you were contemplating a career in forestry, what is the biggest attraction?

3) What skill(s) set is vitally important for any forester?

4) What data are necessary to construct a volume over age curve? Why would you need one?

5) How would you define 'sustainable forestry'?

6) With the various labels extant for aboriginals, Indians, Métis, non-status Indians, First Peoples, and First Nations, is *Candian* a viable label?

7) Do all foresters cut down trees?

8) What makes forest fire fighting an expensive exercise at times?

9) Can forest technologists move on to be professional foresters?

10) In Saskatchewan, what was the downside to forestry Crowns?

11) What could be the single most expensive component in forest fire suppression?

12) Should management be a part of forest fire simulation exercises?

13) With regard to public forest administration, what attributes does a forestry commission format have, as opposed to today's government department format?

14) Do you see benefits accruing from a change in label from 'Indians' to '*Candians*'?

15) Do you see benefits accruing to *First Candian* entities from the introduction of sub-treaties? Would they mitigate the damage done to *Candian* culture brought about by the imposition of Natural Resource Transfer Agreements in 1930?

16) Add a question.

And so forth.

www.ingramcontent.com/pod-product-compliance
Lightning Source LLC
Chambersburg PA
CBHW070849180526
45168CB00005B/1747